Reproduction and Animal Health

Reproduction and Animal Health

Charles Walters
Gearld Fry

Acres U.S.A.
Austin, Texas

Reproduction and Animal Health

The information in this book is true and complete to the best of our knowledge. Because the publisher cannot control field conditions and how this information is utilized, we disclaim any liability.

Acres U.S.A.
P.O. Box 91299
Austin, Texas 78709 U.S.A.
(512) 892-4400 • fax (512) 892-4448
info@acresusa.com • www.acresusa.com

Printed in the United States of America

Publisher's Cataloging-in-Publication

Walters, Charles, 1926-
Reproduction and animal health / Charles Walters and Gearld Fry.
Austin, TX, ACRES U.S.A., 2003.

xviii, 222 pp., 23 cm., charts, tables.
Includes index.

ISBN: 0-911311-76-9

1. Cattle — Reproduction. 2. Cattle — Breeding. 3. Cattle — Fertility. 4. Cattle — Feeds and feeding. 5. Sustainable Agriculture. I. Walters, Charles, 1926- II. Gearld Fry, 1941- III. Title.

SF201.W35 2003 636.2/082

Dedication

This book is dedicated to the memory of James Drayson and Jan Bonsma, two giants on whose shoulders the cattle industry must stand.

Gearld Fry

Table of Contents

Foreword

The man who came to our farm with his stallion aboard a home-made trailer asked my dad, "What do you want, Charlie? A mare or a horse?" By horse he meant a male colt. I was no more than eight years old at the time. Yet reproduction of farm animals was no mystery. At eight you were old enough to lead a cow to a neighbor's lot to have her serviced.

The arrival of that rough looking fellow with two days' growth of beard and his meddlesome stallions brought new knowledge and insight. It was the beginning of an education that was to culminate when Gearld Fry made his appearance nearly 65 years later.

Dad wanted a mare, a request the Ness County, Kansas breeder proceeded to answer with then current, now forgotten, alchemy. To gain the conception resulting in a stallion, the mare to be bred was flushed with a solution of vinegar diluted to artful refinement. A female colt called for a pre-intromission flush with an alkali solution, baking soda. Dad had his explanation. "There are eggs in there," he said. "It is the job of the stallion to fertilize so we can get a colt."

Vinegar is slightly acid. It takes acidity at the point of conception to start a male life. To get a female you have to have a mildly alkali solution for the flush.

All the above is a recalled quotation and can't possibly reflect what he actually said because he spoke in German, and I knew very little English at the time.

Farm life and the wonderment of the life miracle were left behind for the next many years while war and education and editorial work on *Veterinary Medicine* magazine intervened. During my stint at Creighton University, Omaha, Nebraska, I often worked the ambulance watch at Ak-Sar-Ben racetrack. It was heady business learning how to read a racing form, visiting with jockeys and owners, and learning about thoroughbreds. One and all, the horsemen agreed that genetics determine the winner.

Not so clear was the opinion of speculators as to why Secretariat's progeny often failed to express his heritage in the winner's circle. When artificial insemination came along, the Jockey Club decided that this new science had no place in thoroughbred horse breeding. Intromission with a stallion atop a mare must be witnessed and certified to retain thoroughbred status, the stallion having been prepared for his holy task by a teaser mare. After that scientific attempt at progeny management data are archived.

It surfaced again while writing *Eco-Farm, An Acres U.S.A. Primer,* but it was not until I met Gearld Fry that graduate lessons in breeding became available, causing the lessons of childhood to fade into folklore and new science to appear. This then is his story — and mine — and it is presented here exactly as I found it.

Gearld Fry is a credentialed ultra-sound analyst and bovine consultant specializing in reproduction and genetics. Fry knows cattle and he knows how to turn a cattle operation into an economically viable enterprise. He communicates the views expressed in this book from the podium, in the field and across the kitchen tables of cattle producers from Florida and Maine to California and Washington. His signal words come from farmer

John W. Springer, past president of the National Livestock Association, 1898-1903. "There is a smaller hemisphere for the uneducated man every decade and the diminishing possibility of success for the man who does not read. The reading man is in the saddle. The thinking man is guiding our national destiny."

Because of what cowmen read and imitate, they have been lured into the hybrid vigor syndrome. The big cow that is late in maturing and produces pounds without quality is undoing the cattle industry, according to Fry. Cowmen who do their own thinking will be able to see their way through the current chaos.

Fry is a cowman, a consultant and a sparkling new voice amid the bafflement that is rapidly compromising red meat protein production in the United States Headquarters for Fry's Bovine Engineering Consulting firm is in Rose Bud, Arkansas. The arena for operation is the entire country, no area excepted. At one time Fry operated a 500 head beef cattle farm. Lessons learned have now expanded themselves into a consulting clientele that hopes to reinvent beef production.

In western Kansas we had a name for bad looking, scrubby cattle, the kind that exacerbated exotic diseases and yielded poorly at slaughter. They were Okie cattle, always of uncertain ancestry with little or no attention being paid to the line breeding of the ancestors, and with the idea afloat that any male and any female would sort out the genetics needed for herd improvement.

Farmers of the late 19th and early 20th century often knew a great deal about nature. The seven volumes that detail the life of the Randleigh Farm in New York tell more than most college texts about nature's rhythms and requirements. Growing and reproduction seasons, it seems, kick open the door to new knowledge. The late Carey Reams said it all and said it well.

"Life is electrical." All life depends on a supply of electrical energy. Cells are batteries that deliver current. As if by osmosis this fact delivers a second principle, namely that all life alters its environment for its benefit, thus to live and reproduce. It was this cell my dad spoke about without benefit of the nomenclature. Thus the use of the word egg, ova in terms of academia language. The largest cell on planet Earth is the ostrich egg. There are bacteria that hold the title of being the smallest cell. In the horse, the cow, the goat — in all warm-blooded animals — the reproductive cells are the largest and the smallest, the sperm cell being the smallest, the egg cell being the biggest. These cells carry on life's premier purpose, define progeny, establish breeds and preserve lives.

Fry came to his expertise in bovine reproduction quite naturally. At age 20, he and his wife Marjorie bought a dairy farm. There wasn't a bull with that herd of cows, and Fry didn't have money to buy one, entering into the dairy business having exhausted all resources. He concluded he'd have to artificially breed. "I taught myself how to artificially breed," he recalls with pardonable pride. "That was the most intriguing thing I had ever done. It challenged my mind and my thought processes as never before."

Artificial breeding, preg checking, and all reproduction topics consumed Fry's attention, the printed word being the initial tutor. Embryo work followed, as did collecting and processing semen. Not lost was fertility checking of bulls. Many bulls would be 50 to 60 percent live. This asked the question, why? Fry set out to find the answer. He graduated to blood work, using a laboratory in Iowa. "We took blood from animals, dehydrated it and rehydrated it with reagents. That is how we discovered the liver content."

A pixel at a time, a picture emerged, revealing why the semen quality was so low. The results from those bulls became an art, the purpose of which was to bring them up to 70 and 75 percent live with the capacity to freeze semen the result.

The mystery of why those bulls hovered around 70 to 75 percent live semen haunted Fry. As years went by, Fry discovered a font of knowledge seemingly overlooked by academia. There were old timers who had generations of experience locked up in their tenure as cowmen. One was James Drayson, whom you will meet in these pages. Drayson answered many of Fry's questions.

Drayson told Fry it all had to do with the phenotypical makeup of the scrotum. He identified a dimension those testicles had to comply with in order to produce the highest quality semen, nutrition and other considerations being accounted for. The defects of testicles that were too large or too small, Drayson told Fry, could not be corrected. They were genetic. Fry found that only one tenth of one percent of bulls in service were suitable for genetic improvement and peak performance. "For every 50 bulls I semen test, there may be two or three that will comply with the norm I seek, 80% or better live." Often high potency is a single trait, the bull being otherwise out of balance. With high potency at 1 or 2 percent, and all-around excellence at only one-tenth of 1%, Fry concluded that the possibility of creation of the 90 percent bull was near absolute zero as an art of husbandry. A bit of blue ribbons does not settle cows. And association records preserve at least as much wishful thinking as reality, record keeping — like accounting — being ephemeral and errant.

The names for distorted accounting principles are many, and go beyond breeding and phenotype management. They take on names such as agribusiness, vertical integration, factory in the field, confinement feeding. They carry slogans such as the family farm is as obsolete as the horse and buggy. And if we go to grass-based cattle production, millions will starve. And they illustrate how thoroughly informed self-interest can lead to fawning ignorance. Former Secretary of Agriculture Earl L. Butz often told his audiences that 50 million people would starve if natural systems were invoked, if cows were fed on grass, and soil systems were managed without reliance on toxic genetic chemicals and overloads of phosphorus and other salt fertilizers. Butz defended his equation when organic folks stated objections to the presence of hydrocarbon rings in their food web.

A platform speaker has advantages. Butz was seldom asked to explain the logic he used to arrive at his conclusion. One of my veterinarian associates nevertheless asked for an answer and got it.

"Fifty years ago," wrote Butz in a letter, "we had almost entirely an organic agriculture. We used very few chemical fertilizers, practically no herbicides, insecticides or antibiotics. We had half the population. We farmed some 40 million more acres for cultivation. Our exports were one-fourth less." This statement made in the early 1970's — "We had 40 percent of our people actively engaged in agriculture. That's the production record we had with a strictly organic agriculture. Since that time a very substantial infusion of science and chemicals, of pesticides, of antibiotics, of growth regulators and of a whole host of related products have made it possible for us to feed twice as many people at a higher dietary level on fewer acres and quadruple our level of exports. In view of this, the estimate of 50 million people would starve is indeed very conservative."

Thus by semantic sleight of hand, the obviously archaic technology of yesteryear becomes organic farming. That the ability to verbalize does not bestow an ability to think should be at once apparent.

Everyone knows that many aspects of yesteryear's technology were really inefficient soil mining, as indeed chemical technology is today. Ranchers understand, if bureau people do not, that scientific pasture management has very little to do with archaic technology, and that many row crop farmers invoke biologically correct methods to produce more bins and bushels and better quality as well. Cow-calf operators who use rotational grazing and manage the nutrient load in their forage deliver more low maintenance animals to market than do their short-cut blundering amateur counterparts.

All the above explains the Butz "50 million people will starve" statement. Cowmen who think for themselves get a bit nervous when they read and hear such statements. They know the concept is ballyhoo stuff for big industry, big farm technology and the use of millions of tons of toxic genetic chemicals and Rumensin in the bunkers of feedlots and hormone ear tags for the few months of a cow's tortured life.

Does a farmer really feed four or five dozen people? Let logic take over for a moment. For one person to produce that much food, or the raw material for food, by himself, with no outside help, he or she would first need to find a clam shell to fasten to a stick as a hoe or have pastures that permit free-roaming chickens or beef and the stamina of Hercules to do the processing. As a matter of fact, one row crop farmer or cow-calf producer and his family, along with an army of people who produce the equipment, supplies, transportation, information, etc., operate land that brings forth raw materials of questionable quality which is processed into food of questionable quality for those four or five dozen people. Were one to apply some finely tuned economics to the old horse-type family farm operation as compared to modern products and all the ramifications, the horse farmer would look pretty good, even if he also mined the soil. Were the real cost of fossil fuel technology — namely wars and subsidies — to be computed, it would be revealed that we have the most inefficient agriculture in the world.

In terms of a common denominator even grade school students can understand it takes 6.5 calories of fossil fuel energy to produce a single calorie of food energy. In terms of honest accounting, according to a former Michigan State University ecologist William Cooper, "Agriculture ought to function as a closed system. Minerals ought not to be permanently exported off the soil. Animal and human waste need to be added back into the soil system." In the process of production, carbon dioxide, water and energy become part of the food to be eaten and digested by men and animals. Energy is a gift of the land. Water and carbon dioxide — and many minerals — are drawn from the atmosphere. When it functions properly, the system has a caloric balance. But in America it takes more calories to grow, harvest and transport a crop than the crop is worth. Anyone constructing an exchange equation simply has to think of worth as balance. But the "50 million people will starve" advertisers say differently. For 50 years running, they have found justification for farmers farming the way they do, both crop and livestock, all because crude oil has been cheaper than soybean and cotton seed oil. Because of this distortion in accounting principles, modern farming has come to depend on fossil fuels, coal, oil and other hydrocarbons, all fixed in their supply by green plants that lived some 300 million years ago. The technology now sighted as high tech is dependent on ancient plant production.

These few thoughts background our consideration of the cow and the bull as low maintenance animals capable of producing a healthy bull line.

This book started out to be an exercise of linear measuring, then like Topsy it just "growed." The statement that follows will be amplified in almost all chapters, the afterword included. For now, let it stand thus:

Linear measuring is designed to help the producer choose the body type or form (phenotype) of bull and cow that will be high in reproduction and low in maintenance. Measuring of the many different body parts allows the operator to recognize structural and functional defects, which are genetic defects, and potential problems that arise from improper form and type from breeding practices. Choosing the proper phenotype and mating those cows and bulls with each other will build a concentrated gene pool that is consistent and works best on grass. Steers and heifers will finish in 16 to 20 months.

Linear measuring allows the operator to choose the body type or form — phenotype — that the environment around the animals calls for, weather, forage, management, etc.

Linear measuring works as well for the dairy operator. The cows and bulls built on the basis of linear measuring will perform best on grass, be high in reproduction and low in maintenance. Building herds of cows and bulls with the proper phenotype sets the stage for calving ease and longevity. These practices create grass-based genetics. This type of cow must work for the producer ten months of the year or she will become fat on grass and therefore reproductively unsound.

As these lessons enlarged themselves, they prompted Gearld Fry to take a look at pastures, the mineral nutrient load and its effects on both, and finally at the industry itself.

I will not detain the reader with biography, credentials and proofs of recognition. These will surface soon enough, as Fry's art of breeding takes form, gathers speed and moves ahead.

A single reference might prepare the reader for the pages that follow.

In talks with Fry about reproduction management I wrote down the notes that became the pages of this book. I used the personal pronoun "I," and it became difficult to know I, Fry, from I, Walters. Therefore let it be understood that "I," this writer, appears no longer in the pages that follow, except in the Afterword. Starting with Chapter 1, "I" means Gearld Fry, and the texture of rhetoric leans on his knowledge and comprehension of the breeding art, the role of the wordsmith being that of a craftsman causing the narrative to flow, eddy, swirl and do for the message what Fry and ancestors have done for the development of sires and cows capable of sustaining improved herds, superior meat production, and finally, human beings capable of thought and reason.

Reproduction and Animal Health is more than a story of *Bos taurus* and *Bos indicus,* the two remaining genus and species of bovine on a planet that once had a dozen genus and species. It deals with the affection one had for Bossie, the family cow, and with the art and science of husbandry. It indicts corn as a feed not suited to this herbivore, and opts for grass, the metaphor for an assembly of plants often called the forgiveness of nature. The great Kansas State Senator John J. Ingalls called grass "nature's benediction." Finally, *Reproduction and Animal Health* pleads for humane treatment of the bovine and a return to pasture.

— *Charles Walters*

Chapter 1

Reflections of a Cowman

It has been 20 years since I built the reproductive center at Rose Bud, Arkansas. The center was simply a prefab building and a lot of pens to hold bulls. I had available pastures to maintain cows for customers who wanted them flushed. I managed cows and bulls for clients. I sold that business to our older son who now deals with fertility management of horses.

When I sold my business because of a temporary health problem, I purchased ultrasound equipment. I knew there were problems in the industry, and I wanted to help solve them. I figured that with that ultrasound equipment, I could evaluate carcasses and make recommendations — I'd fill a void. I had in my mind as to what was wrong with the cattle business.

I very quickly found out that it was not a valid tool for my objective. All it did was provide a living. It did nothing for cattlemen.

I ran into an old gentleman named Harlan Discus in Houston, Texas, several years ago. He had a bull worthy of Michelangelo. I observed that bull for 15 minutes. I had never seen a sire with that volume of meat and with the balance of that animal.

The old gentleman said, "What do you think, son?" I sat down and talked with Harlan for three hours. I repeated that seance the next day. He told me about linear measurement. He told me how he selected animals, what he had learned and how he did it. He invited me to his place. I went to Faith, Nebraska and spent a week with Harlan, and I still tap into his knowledge with questions.

That was my first encounter with linear measuring. It has opened a world of knowledge I now hope to share.

Harlan Discus, Dr. Jan Bonsma and Dr. Burl Winchester were friends. Bonsma visited America in 1960. In 1963 he returned from his native South Africa and met with Dr. Burl Winchester at Montana State University. He introduced Winchester to linear measurement. Linear measurement was researched at that university for a number of years. When the research was complete, Winchester retired and formed an association with Discus, a man by the name of Innan, and Carney Redman. Nebraska University also developed a database, then promptly archived it into some inaccessible ward. Winchester and his friends lectured and taught. Their seminars taught cattlemen to establish better herds.

That was in the 1970's. It was at the height of import invasion by Charolais, Limousin, Simmental, and the other continental and exotic breeds. Cattlemen were crossbreeding and outcrossing, usually with mixed success. Tall slab-sided cattle were being created. This theory period was being touted by university studies and was made to suggest the wave of the future.

Winchester and his associates had closed down their organization when I met Harlan Discus in Houston and he showed me what he could do. I was off and running. I knew I had discovered the break in the dam. It has been a learning process ever since.

Linear measurement is no come-lately idea. Archives in Europe reveal that they were linear measuring in the 15th century. They have on file the basics that were recently made a matter of record by the South African Jan Bonsma.

This much stated, we have to wonder about why the first real attempt at making the scientific system in cattle maintenance settled on EPD, expected progeny difference, rather than linear measurement. The former merely enables single trait selection, whereas the latter delivers to farmers the tool for eliminating the guesswork from judging and removes the university and the breed organization from control.

Keeping the sanctioned body type in place complies with public policy, namely the industrial model for grain production, the transfer of energy from field to feedlot using grain that is subsidized to the mega-grower, world priced to the feedlot, the resultant product questionable and dangerous to the public health because of hormone overloads and daily diets of antibiotics. It can be said with little fear of contradiction that the dairy industry would die promptly if cheap corn was not available.

The system that is in place does not permit the cattleman to earn a return on his investment and work. It has created an animal it costs too much to keep.

The aim of science is to foresee. This is the reason records are kept and databases are constructed.

My good friend Darol Dickinson has researched what I now relate. It carries a jolting message that plugs into our persistent problem.

Jump Starting Genetics

The dictionary defines *science* as "a branch of study concerned with observation and classification of facts, and especially with the establishment of defendable general laws."

The breeding of cattle on a worldwide basis swings the pendulum from one extreme to the other. In 1880, Texas Longhorns were rounded up by the millions. Certain bulls were castrated, and others were retained to breed the herd. If a bull calf was fleshy, muscular and appeared to be of a good quality, he was castrated to develop into a saleable steer. If the bull was of low quality, poor doing or small he was left as a bull, later to be a herd sire. This early Texas practice, described in J. Frank Dobie's book, *The Longhorns* (1939), was popular when the West was unfenced and millions of cattle reproduced with no economic thought of insurance, interest or taxes.

The totally opposite example happened in Germany starting right after World War II. The country was devastated by military destruction and ravaged by disease. People had TB and so did the cattle. Hoof and Mouth disease was only one of numerous problems with no CNN reporters present, or even caring. The German national cattle herds had been consumed by starving soldiers and surviving residents. Very few farm livestock remained. General George Marshall, under directives of Harry Truman, developed and implemented the Marshall Plan designed to use U.S. funds and leadership to rebuild Germany and the other European countries destroyed during the war.

Heino Messerschmidt was the German undersecretary of agriculture. At Neustadt Aisch the Besammungs Verein genetic program was developed. The superior genetics of Germany and Europe were pooled for early artificial insemination use. There was an all out program developed to eliminate disease transmitted by natural copulation, and provide the highest scientific proven genetics to poor and starving people. Semen was processed on beef cattle, draft cattle, milk cattle, milk goats, milk sheep and swine. At this time artificial insemination was not a frozen product, but hot. There were no plastic AI sleeves.

Each mating was important — very important. Families were living on milk and meat. Disease could not be tolerated any more. The war was over.

Heino Messerschmidt was Cambridge educated. His job was to reestablish the German genetics, then develop them to be the greatest in the world. The mentality of the German people was to do whatever it took, pay whatever price it cost, and become the best.

Every male admitted at Besammungs Verein was performance tested. Only one in thousands became a useable semen sire. It became a regulation that no one could own his own bull.

No one swayed by personal feelings, sentimental love or even economics was allowed to own a male, or to breed to anything except government proven, tested males. All breedings were AI, a beef specialist individually inspected every cow and approved each mating with careful calculations. It was a regulation. (In Germany a regulation is as good as a law.)

World-renowned cattle scientist Jan Bonsma of Pretoria, South Africa was associated with Messerschmidt to develop the ultimate performance guidelines. He left no stone unturned to define the superior genetics.

The Besammungs Verein stud farm made the world's fastest gene-tic advances. They have been testing for over 50 years. (The Gelbvieh breed is one result of their efforts.) It is now the world's largest semen center. The first calf born from frozen embryo technology was at Besammungs Verein.

From a meek beginning German farmers with one to five cows have developed sires of world-renowned quality. During Messerschmidt's lifetime, genetic improvement happened so fast semen and sires have been exported to North America and worldwide for herd improvement. Today the livestock quality in Germany is of world leadership vintage and superior to genetics from areas never touched by the ravages of war.

What was the success formula?

1. Identify superior sires.
2. Performance test large numbers.
3. Do not allow natural service breeding to inferior untested bulls.
4. Do 100 percent AI with superior proven genetics.
5. Provide professional assistance for every mating.
6. Create genetics so superior the whole world wants them.

In the United States there are no government regulations governing the purchase of semen. No professional government regulator will knock on your door to see if there is a cull bull with your cows. These are things you will have to enforce yourself.

But the result achieved by a serious postwar U.S./German effort of rapid breed improvement, profitable production and serious quality control is available to everyone who will commit to the same success process.

Line Breeding

The horse that was brought to Iceland from Norway in the 9th or 10th century is both a symbol of national identity and the epitome of line breeding. The animal is small, has a heavy coat and preserves a five-gait style for travel. It is more surefooted than a Grand Canyon burro. It endures a climate too harsh to permit the maintenance of a dairy cow. Its winter coat has the earmarks of an electric blanket. The animal is well adapted to its environment. Iceland has prohibited the importation of any other breed, crossbreeding being considered as some sort of insanity.

Ishestar is a riding center near Reykjavik, a policy maker, if you will, that prescribes and proscribes. Importation of strange breeds, it is reasoned, would destroy the uniqueness of the Icelandic horse. Instead of hybrid vigor, the race would degenerate into mongrel status, forever an incubator and retailer of diseases that have never gained a toehold into Iceland.

Many centuries of breeding the best to the best have delivered a phenotype characterized by five gaits — a sideways trot, a type of running called tolking in which the rider feels little if any motion, and the three gaits common to all horses.

It has been written that the area now known as the Black Sea was once a below-sea-level Garden of Eden. When melting ice caps created enough water to tear out a dam at Bosporus Strait, the Great Flood of Gilgamesh or Noah followed. The population dispersed from the basin that is now the Black Sea, and with it went domesticated animals. The Tartars of Asia were the first to gentle the horse for riding, for farm work and for war. Gene pools that produced the first animals were not shared, according to biblical tradition.

The great sagas tell of wondrous deeds, of trials between man and man, beast and beast. *The Story of Burnt Njal,* translated by George Webbe Dasent, tells of two stallions fed and trained to fight to the death for heavy wagers in 10th century Iceland. This terrible fight resulted in a blood feud ending in a massacre.

The Spanish cattle, now known as Longhorns, were line bred by nature's necessity. Harsh climate and brutal droughts demanded the survival of the fittest in the herds, so that now the breed retains one of the best genetic potentials for further development.

Most of America's surviving farmers raise cattle. The public prints call cow pen keepers cattlemen, and public policy makers applaud the opponent of family farmers on grounds that economic concentration will make meat protein cheaper. Nevertheless the greatest common denominator in farming is cattle. Farm numbers continue to sink, yet a recent family farm inventory notes that most of the beef is birthed and grown by most of the remaining family farmers. Still the cattle business is the most concentrated of all farm activities. No more than two large packers kill, gut and market close to 80 percent of all the slaughtered cattle in the United States As cattle producers absorb losses, the monopolies sees profits grow exponentially. As farm numbers decline, agribusiness grows. One or two firms feed most of the fat cattle in the pens after farmers have handled the cow-calf and backgrounding end of the business. They have been assigned a position of dominance with the blessings of public policy and farm organizations doing nothing about it. The intellectual advisers of agriculture have applauded each step, each maneuver, each consolidation until I am forced to conclude that the highest aspiration in academia, in Extension, in USDA itself, is the annihilation of the farmer clientele.

The bottom line to all this is commodity beef, tasteless beef, and bad eating experiences by those tasting meat protein fare in five-star restaurants at least two out of ten meals. Commodity beef with little or no attention to sound breeding procedures not only destabilizes human metabolism, it is indicted for the destabilization of the economy as well.

The cow-calf operator is not exempt from the ruin designed for him. He has been lured into what I call the hybrid vigor syndrome. The big cow is late maturing and provides pounds without quality. It has become almost impossible to manage and support this animal at a profit. The seed stock providers have forgotten how to produce a bull commercial cattlemen can use with confidence. Fully 85 percent of a bull's progeny should qualify as replacement stock or be fast gainers that grade choice or better. Some 50 years ago 75 percent of fat cattle graded choice or better. Today the figure is near 30 percent. As a consequence, the meat protein industry constantly laces meat with chemicals to supply taste. Yet this is the responsibility of the producer on the ranch. Meat should not require these enhancers. Quality stands on its own. It is this faltering quality that has consumers drifting away from red meat. And it is the feedlot's rape of the cow-calf operator that is annihilating the gene pool.

As with hybrid corn, the industry has created an animal unable to supply a suitable level of testosterone and estrogen for the early development of muscle pattern. It takes compliance with nature's rules to produce a carcass with quality. But this does not happen when schoolmen refuse to transfer knowledge to the farm or when public policy allows debased technology to run rampant over the health of the people.

Two Systems

There are two systems of thought before the farmer. One leads to marginal productivity, degeneration, mongrelization and penury. The other defines the role of the farmer-entrepreneur, sound husbandry, quality meat protein and cow operation solvency.

For well over a half century, the collective wisdom of scientific breeders has been invoked, the new science chanting hybrid vigor, hybrid vigor. Yet the evidence of centuries argues that creative evolution requires a line program faintly reminiscent of biblical genealogy.

Blunders have a way of repeating themselves. At the beginning of Century 21, the breed association strategy would ratify what has already happened. Thus, the seed stock producer is facing the worst of times and the best of times. The Black Angus breed has been one of the benchmark beef animals for fully a century. Much of the breed's status has been accomplished by seed stock producers who have resisted rampant mongrelization. Early in Century 21, the Black Angus Association leadership voted to share the bloodlines and track progeny welfare with databases of ancestral records and genetic predictions of Angus derivatives. This decision seems to say that the commodification of beef is both a *fait accompli* and an opportunity for a branded product based on the Angus phenotype developed through the years.

This intelligence tracks the old EPD — expected progeny difference — idea that has dominated and doomed beef quality. EPD selection, not line breeding, is to be the wave of the future.

I shake my head with abject disbelief. They've taken the wrong road. The right road, implied, requires psychoanalysis, a rehearsal of events that brought the industry to its present impasse. The show business part of cattle raising has long dominated breed management, often permitting recessive genes into the breed, as was the case with dwarfism in the Hereford herds of over 60 years ago.

Every breed, it seems, has a charming history. No great purpose would be served by a mini-encyclopedia codification at this time. It is what has happened to herds during the last half century that defines or preserves the future, whatever that may be.

British Whites

The British White breed came to Scandinavia in the eighth or ninth century, origin unknown. After transfer from Scandinavia to England, the integrity of the breed has been almost a carbon copy of breed maintenance for the Icelandic horse. Historians recalled that a polled white cow with black ears grazed the mountains of Scandinavia at that time. A few were brought to the British island by Vikings, either for trade or as conquering settlers, no one can be certain. The meandering of herds through the park of Whalley Abbey, later in the forest of Boland, Lancashire, finally became the private property of royalty.

During World War II, Winston Churchill ordered British White cows and a bull shipped to the United States — in case. In case the Nazi war machine overran England, total annihilation of the breed was to be expected, probably because of the total disdain Hitler had for British royalty. British Whites are not to be confused with White Park, a breed that has lyre-shaped horns and its own registry.

Five cows and one bull of the British White breed were shipped to a Pennsylvania prison farm. Those five cows and one bull became the foundation stock for the British White breed in the United States. Today the British White is a holdout against the EPD system that links to government genetic advances with a computerized registry.

Without adequate records, no one can say whether minor genetic influences were or were not introduced into the British White. Nevertheless a penitentiary bull line headed by an animal named Old Ugly accounted for Atlas, and records now identify the Old Ugly-Atlas genes in British White herds. Imported bulls direct from England have helped British Whites hold the line against finding seed stock via hybrid attempts.

Tom Zimmerman of Des Moines, Iowa imported Wood Baskwich for a herd since sold to the Jim Kelly ranch in Minnesota. British White is not the only valid breed, but it serves a good model for turning the industry around.

After all, the well bred cow will give you a live calf year after year, more pounds gain per acre of forage, with assistance at calving practically non-existent, even heifers with their first calf.

There are signs that enable cowmen to read an animal like a fingerprint. These are measurements that foretell, most of which are clipped away in deference to the show ring. Color does not necessarily relate. For instance, blood protein studies have revealed that the horned, black-eared White Park and the polled black-eared British White have no blood relationships whatsoever, no more than do the Angus, Hereford, Shorthorn or other British breeds. There are no less than eight breeds with colored parts listed in *Cattle: A Handbook to Breeds of the World.*

The British White is genetically hornless, docile, and was originally a dual-purpose breed until about 1950. Since then it has been "all beef." "Trouble free," says Zimmerman. "That's a British White."

Each breed has its defenders and promoters. I hope this handbook takes cattle breeding by the nape of the neck and the seat of the pants and shakes out some very enlightening results.

Up front, I opine that the major problem with open cows is the low quality of bulls. The problem seemed intractable a few years ago, but in recent years the mood has changed.

Bio-Correct Farming

Ranchers now see their role in the bio-correct farming movement, in keeping cows on forage, in dropping a different kind of calf. Still keeping cows on grass is possible with the right phenotype genetics.

I hold that the seed stock producers do not have the bull the primary producer needs. Moreover, they cannot answer the questions even a non-cowman journalist might ask. The exploiter has been allowed to be the economic counselor of the cow-calf operator. Much of this book will focus on the bull, and the bull that I can hang my hat on isn't out there.

James Drayson of Big Sky Genetics, Billings, Montana, tested bulls for 35 years. He followed 1,500 bulls from birth to death and recorded everything that happened to each animal. He measured them at least twice a year, took semen, analyzed it, tracked progeny, and recorded his findings in a monumental work entitled *Herd Bull Fertility.* He set standards for young bulls no more than six months of age because he knew what type of sire the animal would be at maturity. He categorized each bull, the details of which will appear in Chapter 4.

A balanced nutrition level has to be available, and this will be covered in a chapter called *The Mineral Diet.* Normal sperm cells are not possible without a proper ration. Without attention to nutrition, high abnormalities, broken tails, low mobility — things that keep a bull from settling cows — become the producer's legacy.

A recent influx of new persona into the cattle business was attended by neither folklore nor a scholar's knowledge of genetics. New owners of spreads and cows knew nothing about mixing and matching bulls and cows. The exotics made their way into America. They were different. They tracked excitement in the showring and they supported crossbreeding and downgraded line breeding as an incestuous procedure. Husbandry vanished. In veterinary medicine annihilation of the beast became the cure for disease, a case of dealing with symptoms, not causes.

Yet to survive in the cattle industry the cow-calf producers have to use bulls that are masculine and pre-potent. They must produce cattle that are efficient and adapted to each individual environment. We're fools about moving cattle across the United States.

This won't work very successfully. Cattle born and raised in New Mexico or Montana should stay there. If they are born and raised on fescue, they should stay in their environment. Transport to distant pastures usually means loss of a year's production somewhere along the line. It's up to the cowman to produce genetics that provide a uniform, consistent red meat product that meets or exceeds consumer demand. Cattle growers who wouldn't read this book usually don't care what kind of product they're producing. Their product is a commodity, one not distinguishable from the rest.

There is an attitude afloat that goes like this . . . I want a bull that will give me the most pounds. But the faster you grow it, the tougher it is, the sorrier the meat. Those five- and six-pound gains a day put an unpredictable product on the other side of the equal sign.

Blunders A' Plenty

Overstocking, overgrazing, poor reproduction efficiency and lack of marketing savvy all conspire to threaten the producer with insolvency.

As I sketched the outline to be followed in writing this book, I wanted to drop these verbal bombs in clusters and in haymaker payloads.

Item. Fifty percent of the total energy used in feeding a pound of beef is used for maintenance of the animal. We can do better.

Item. Increasing milk production will increase the maintenance required of the cow on extended calving intervals. In short, overloads of milk production and reproduction do not go together. Reproduction is more important than price.

Item. According to an Australian study, the reproductive rate is 50 percent of the economic value of the cow-calf farming operation.

Item. Low maintenance and high fertility levels go hand in hand.

Item. The average calf has seven or eight owners. It travels 1,400 miles from birthplace to dinner table. About 30 percent grade choice. At a recent Hereford Association meeting, one speaker pronounced Hereford select as better than Angus choice based on Colorado State research. Such nonsense sends out the message that nothing should be done, nothing needs to be done, and that out production of the Angus is a *fait accompli.*

Item. The average calf has a sheer force of eight — on the Warner-Broad scale — which is tough to eat, yet this index defends feedlot to supermarket fare. Fat means energy. Energy causes cows to get pregnant. The bull is not exempt from the above item. The world is full of college research revealing how to decrease the amount of subcutaneous fat on the bull, yet the seed stock producer who chooses on the basis of this research will "hurt bad," as the old cowhand would put it.

Sometimes the wrongs overwhelm what is right with a herd.

Seed stock producers tend to think in terms of bulls, the prize bull being the epitome of success. The late William A. Albrecht of the University of Missouri had his answer. It was simply that the cow had something to say about the character of the offspring, 50 percent in fact. This asks the question, which heifer out of which cow makes the best replacement? Always in question is single trait selection, as suggested by EPD.

Cowmen with excellent genetics governing their herds refer to the commodity meat as "mule meat," and they assert the stores are full of it.

This reality poses a question. Are there enough genetics out there to turn the industry around? Jan Bonsma in *Man Must Measure* told of his tour of the United States during long absences from South Africa. He examined cattle in most states and concluded, in 1968, "I am afraid the American cattle industry has moved to the point genetically where there's no return" — without outside genetics, that is.

That is not entirely my opinion and the opinion of Allen Williams of the University of Mississippi. We both are more optimistic. We both believe there are bulls and cows which can do what needs to be done.

The Bohaty Ranch

My take on the British White was handed to me during a field day at the Walter and Nancy Bohaty Ranch, Belleview, Nebraska. It was a typical field day for some 50 breeders, genetic and forage experts who enjoyed the pasture walks and on-scene

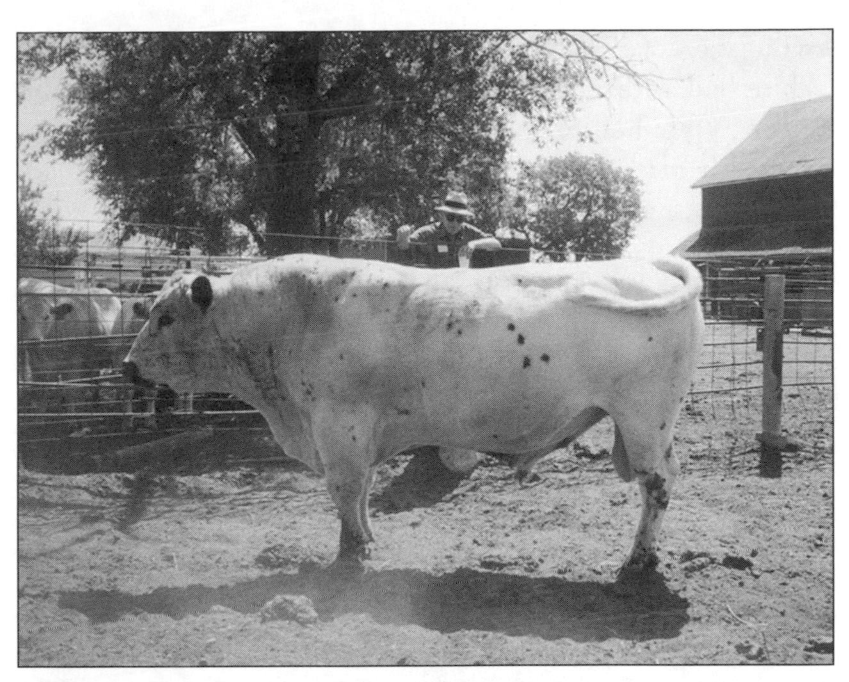

A British White bull (above) with a first cross illustrating how this bull stamped his progeny with his phenotype.

examination of Bohaty's British Whites. A snapshot of a British White bull with a half-blood cow revealed the ability of the British White bull to stamp his progeny. Most British Whites completely mark the phenotype and color patterns of the progeny in the first cross.

The cow pictured on page 15 is the kind and type typical of a first cross and should be a terminal cross and slaughtered. Such a cow is typical of the hybrid vigor resultant from such a cross. The British White bull is an excellent bull for producing terminal cross meat.

Europe is the oldest producer of cattle genetics in the present century, and it continues to have some of the wisest cattle breeders in existence and the highest concentration of genetics in the various gene pools of many herds. I confirm that British Whites have prepotency levels equal to or exceeding most breeds in the U.S. The breed has an average quality grade of 90 percent choice, 17 percent of the 90 percent grading prime, an average yield grade of two or better.

Some parts of my *Reproduction and Animal Health* story surfaced when I joined *Acres U.S.A.* in interviewing Tom Zimmerman, a part of which I now quote:

ACRES U.S.A. Are you able to collect for this differential?

ZIMMERMAN. If they put them on yield and grade, they are, yes. And I think some of the feeders are starting to understand that and starting to do that. You were down visiting Bob Glenn, and I think Bob said that if he had run his cows through on a yield grade, he would have collected about $30 to $35 a head premium on them.

ACRES U.S.A. Isn't this the problem with the cattle business? The producer is too focused on raising a commodity product rather than a quality product?

ZIMMERMAN. I think in a lot of cases that is correct. That's the reason that you get into crossbreeding so heavily, because you do get the hybrid vigor in crossbreeding. It's like the corn business; the average corn farmer really doesn't care about the quality of the corn and the nutritional value of the corn, does he? What he cares about is bins and bushels. That's it, that is all he cares about. The basic cow-calf man, he cares about the number of pounds that go across the scales. That is what he cares about. The industry is beginning to change, and I think we are going to see, in the next 15 or 20 years, quality beginning to be compensated with premium dollars. When that happens, in my opinion, this breed, British White, will come to the forefront.

ACRES U.S.A. That will depend a lot on finding the sires to keep a quality crossbred animal coming across the scale, won't it?

ZIMMERMAN. Absolutely. Take that bull we just saw out in the pasture. I will guarantee anybody that if we could collect 10,000 straws of semen from that bull and send it to a ranch that would artificially inseminate 10,000 cows with it and get 7,000 calves out of them, those would be the best 7,000 calves that they ever ran into the feedlot. I will guarantee that. It's just that the cow people, wrongly or rightly, are very reluctant to change unless they can see it in the auction market. Now, in the last seven or eight years we have seen everything go black. That's because the black calves have been getting a premium. Consequently, Angus bulls are the "in demand" bulls for most commercial cow-calf people right now.

ACRES U.S.A. It's slipping right away from Hereford, isn't it?

ZIMMERMAN. Hereford has been kind of on the decline for several years. In this area, Iowa, Charolais and Angus are the two biggest breeds.

ACRES U.S.A. In order to have a good crossbred product, you have to have a good sire. How can you determine whether or not you are going to have a good sire out of a male calf? Does it require linebreeding entirely?

ZIMMERMAN. I would say yes.

ACRES U.S.A. How would you describe linebreeding?

ZIMMERMAN. Linebreeding is the breeding of a male to a female, regardless of the parentage. You breed the best to the best of a particular line. The breeder has to have a vision, and that vision has to be in front of him as he picks out the male to breed to the female. The breeder has to see out in front, much as an artist sees and looks at a canvas. He has to be able to see that picture on the canvas before he starts painting it in, doesn't he? The cattle breeder has to have the image of that animal there on the canvas before he mates the animals. The cattle business being what it is, it's a long process.

ACRES U.S.A. You mean to set up the right kind of sires?

ZIMMERMAN. Yes, it is a long process; it's not a short process. When this breed (British White) was first introduced into this country, it did not have large hindquarters. It was a dual-purpose breed in England, and as such you didn't have much meat on the hindquarter of the bulls. Consequently, it has been a long process to get bulls with hindquarters like the ones we have here. That has been a 20-year process. And it is a hit-or-miss kind of a deal, because you may get a bull with a great hindquarter on him, but he has a terrible disposition, or he may have a great hindquarter but he has other problems. You don't want to use him because he has other problems. It is a matter of selecting what you think is the best sire to the type of female that you want to use. I am a great believer that the female in your herd, who year after year consistently produces the top calf is the female from which you want to select a herd sire, because you take her genetics and spread them out through the calves from him. If you have three or four or five top females in your herd, then basically you can select a sire that meets our requirements off one of those cows.

Now go to the next chapter. Learn the signs that measurements will confirm, and foresee the days of grazing.

Chapter 2

Reading the Animal

The name of the game is cows and bulls, and how to select. Before I go into linear measurement I want to caution you to see what you look at. If you find the clues the Creator has installed for your inspection, you might find that in time you no longer have to measure all the animals all the time. There are signs you look for physically. In fact the potential for performance is almost engraved in hide and hair.

Selecting Livestock

The following abstract and extract from the work of Jan Bonsma should set the tone for this chapter and those that follow.

The fact that the Creator has made us a little lower than angels and has crowned us with glory and honor is an inspiring position. He has given certain people the responsibilities (gifts) of producing food and fiber, keeping the people of the world clothed, fed and healthy.

This places a tremendous responsibility on man; he is the axis upon which all livestock and agriculture production centers.

Man's responsibility compels him to control his animals, to measure them and to select them, and to manipulate their genetic makeup. When breeders are rewarded for a high-quality product, they will produce without outside interference.

At the moment of conception, the complete genetic potential of the animal is laid down, this function to perform or produce during its entire lifetime. The boundaries are rarely attained however. The environment, including the intrauterine environment, constitutes a severe limiting factor when it is unfavorable. It limits the full manifestation of genetic potential of the animal; it retards skeletal development and ultimate size, affects muscular development, hide thickness, length and appearance of hair coat and the glandular function. All play a vital role in the body profile and various related anatomical features.

Likewise, genetic or hereditary defects or limitations are reflected not only in function, but also in appearance. Of all domestic animals, the bovine lends itself ideally to judgment for functional efficiency. Its conformation features are distinct and can be judged in terms of measurable production and efficiency.

To know the genetics of our livestock and how they function and the performance of those animals is of utmost importance. There is no system in the world (numbers or otherwise) that can mix and match animals without visual appraisal for form and function.

The endocrine function is the most important function of any animal you choose to put back in the herd as replacement, male or female. The hair and many swirls in the different gland locations, the testicles with crest on bull's neck, and ovaries and rump area of the female are the window into the development of the endocrine function, which is from the genetic makeup laid down at the point of conception.

The Endocrine Stimulus and Function

The endocrine system, controlled by the pituitary gland of the bovine, basically controls the conformation, performance and

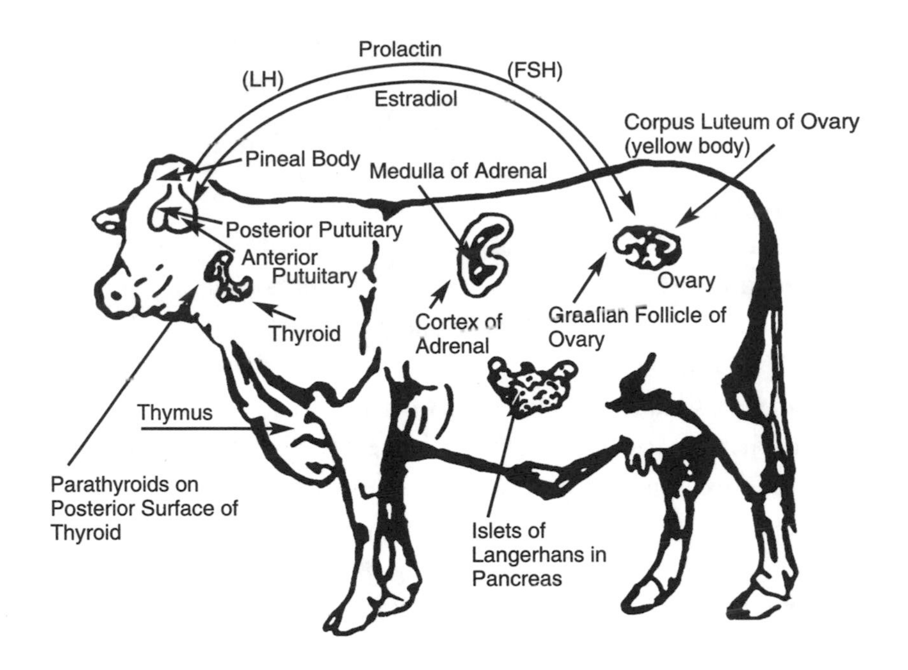

production. It determines the hair coat of the animal and the animal's ability for adaptation and performance. The pituitary gland is in close contact with the brain and reacts quickly to sensory organs, ears, eyes, nose, skin, and feelings. It may be described as the "conductor of the orchestra" that determines between fair, good and excellent performance. The pituitary also regulates the actions of the thyroid, pancreas, thymus, testes, ovaries, adrenal gland and liver. Each of these functions are totally controlled by the genetic makeup of the animal as information is passed to and from the brain. The function is a result of the breeding principles used at the point of conception.

Early development of the pituitary is most important for the proper development of the testes and ovary. The testes produce testosterone and the ovary produce estrogen. Early development of the pituitary creates large testes and ovaries. Proper development of these two glands determines the degree of masculinity and femininity each bovine achieves. The production of hormones in the testes and ovary alone with the genetic makeup determines the amount of red meat an animal will have on its carcass. Proper development of the pituitary determines how quickly the animal will finish on feed for slaughter using grass or grain. Early development has great affect on the phenotype of the animal. Early development is not only responsible, but crucial for early puberty.

Choosing bovines for early development of the gland system and mating them to animals with like systems over two to three generations will begin to concentrate the genetic makeup in your herd. Also, in the progeny produced the function will become a homozygous trait. Characteristics of the third generation calves should have thick hides, short hair, vector resistance both internally and externally, reproductively sound and well balanced.

Late development of the glandular system prolongs the onset of puberty, resulting in slow breeding in cows and bulls, late calving of cows, calving difficulties and inconsistent production. Late or slow development of the pituitary causes bulls to look like females and females to look like steers.

As has been shown, measurement for development and function of the pituitary determines the performance of the animal. The size and shape of the muscles are also determined by pituitary function. The heavier muscled animals have the early maturing pituitary system. The development is all very visual and easily read. The points of measurement, when within the window of balance, are important in selecting for a functional gland system. Measuring the muscle patterns and correlations of an animal allows you to know how the animal will perform.

In the process of measuring you cannot separate measuring the semen quality from measuring the muscle patterns. They are all correlated. If the testes are small at 12 months, the animal in question will never reach a high degree of fertility. The composition of the testes, scrotum and nipples of the male have a direct affect on the udder conformation of the daughters produced. The same is true if the cow has small ovaries or does not begin cycling until after 10-12 months. If late puberty does occur and late development is achieved the animal should be slaughtered.

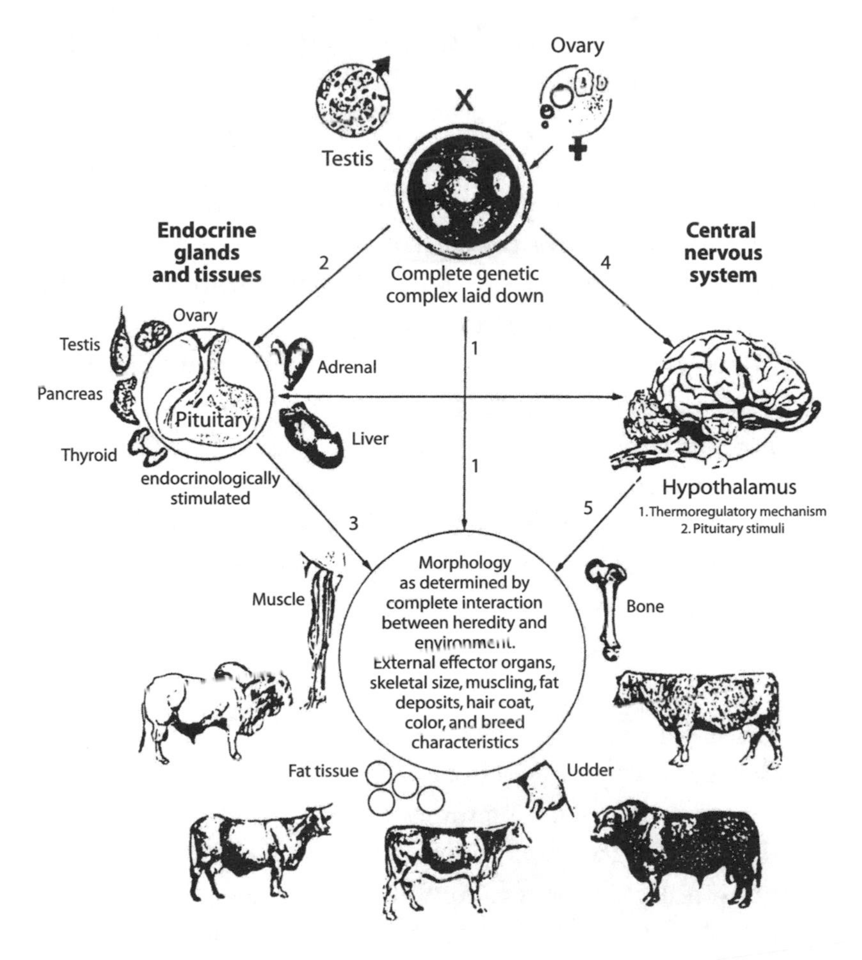

The interaction between genes and the phenotype. At the moment of conception the complete genetic potential of the animal is laid down. This determines irrevocably the potential boundaries within which the individual can function, perform or produce during its lifetime.

As has been stated, hormones are responsible for the volume of muscle development in the bovine. The mixing of genetics, the management of the animals, the ability to provide the proper nutrition (through grass and minerals), the environment and its affect on the animals *are all areas we as herdsmen have the greatest responsibility.* We, as cattle people, have not been taught how to look for structural or functional defects. *The function of the endocrine system is the most important of all.* It provides for taste, palatability and function.

Livestock agriculture is about providing a high quality and healthy food along with a wonderful and palatable eating experience for those who consume our products. Drugs, chemicals, synthetic hormones and crossbreeding to produce pounds does not provide for healthy food. There are cattle in the world that consistently gain 4 pounds a day on grass with no health problems. Grass is free, a God-given raw product which converts to red meat and milk, provides protein and makes for a healthier and more pleasant eating experience.

If you select for reproductive performance and low maintenance, you automatically have the carcass you need. This borders, if it does not encroach, on being an absolute. It cancels out single-trait selection.

EPD Shortfall

This is why EPD will take you down a dead-end road. This so-called scientific approach came out of New Mexico in the 1950s or thereabouts and presumed to set up a database that predicted performance, replacement potential and meat protein quality. In reality the system complied with the scientific finding called single-factor analysis. I know of no cattleman who has the ability to use EPD — expected progeny difference — and not single-trait select.

The hunt is for milk, for hide, for length, for something. There is nothing about EPD that tells you anything about reproduction or low maintenance, the real reason for this discussion.

Admittedly, EPD causes cows to perform on paper, and this translates into sales. I mention this as an aside. I want you to use the mind's eye to view a cow from the front. She should be wide. She'd have a bit of rump. Reproduction in the cow is in the rump. Wide shoulders and a deep chest must compliment each other. Without the deep chest it does not matter how much rump there is. Without that balance you won't have the maintenance to support reproduction.

The Cow

If you have a cow with a carved-out chest girth flanking the front leg, she's a high-maintenance animal. This asks a question and suggests an answer. The great racehorse Secretariat had a heart that was three times the size of regular thoroughbreds. Do cows and bulls that are correctly constructed in terms of chest and rump have an improved circulatory system? Do such animals have a different respiratory system? What are the signs that identify the character and cost of the individual regardless of breed or EPD statistics? I doubt that depth and width tells the whole story. Some philosopher once said, *The curse causeless shall not come!*

Of course, I know why we get that carved-out effect. EPDs and registration certificates say genetics.

The Result

I defy any individual or group to tell me which of the following silhouettes is the larger or smaller, using eyeballs alone.

Your senses seem to lie. In reality, both are the same size. An overview dismisses too many details. A handsome bull with small hidden defects is little more than a big pile of deficits in the bottom of the cash drawer.

The legendary question presents itself, what do you see when you look at a herd? You'll see a cow that's limping. If she's sick she'll have her ear down. You would probably miss it if her eyes were sunk, unless you consciously look at her eyes.

Look at the baby calf. Is a navel cord in evidence? If there is no navel cord there is a problem. The problem with a calf tells you that the mother has problems, a case of one generation visiting its sins on the next. Up front the missing navel cord tells you that you'll have a sick calf in the fullness of time. It will have names — navel ill, diarrhea, pneumonia. A nutritional deficiency causes this problem, and it started with the mother.

It all feeds back to poor-quality soil, grass, hay, and the availability or absence of minerals in the feed ration. Some problems are flexible, some not. The last are locked in the genetics, the DNA, and they can't be whisked away with certificates, regulations and spin.

Problems

Foot problems, pink eye, retained placenta, cattle that don't breed, calves that are lethargic, stillbirths, hair that does not shed — all these observable signs have answers — and questions. Some parts of the questions and their answers have to do with genetics and some call management practices on the carpet.

Two Outside Signs

The outside signs of inside genetics ask for attention before we enter the arena of linear measuring.

The glandular system of the bovine is key to comprehending the selection and reproduction role. It has to function. It's the motor.

Admittedly, the bull has to be good enough to lend himself to a cow so that she can express herself. There is a process known as cytoplasmic inheritance. Seated in the unfertilized ovum is genetic material that is not affected by the sperm cell. Cows that are made right exhibit a body form that comes from the X chromosome. The cow has two, the bull only one. That is what I mean when I say the bull has to lend himself to the dam. The role called for is filled by a bull without recessive genes. Recessive genes tear down and destroy. If the female is prevented from expressing the quality she has to her offspring the reproductive system is short circuited.

A Breeding Program

A breeding program should focus on cows, not bulls. It breaks down further into meat and economics. The goal is to breed a cow and have her wean a calf by the time she is 1,000 days old. That means she has to be bred by the time she's 14 to 16 months of age. The cow that calves at two years of age by the time she's 1,000 days old will have produced a half pound a day.

The cow that doesn't calve until she is three years old — 1,365 days — has just produced 40/100 of a pound a day. To get that early calf requires the right phenotype and a pelvis capable of delivery.

All things are determined at the point of conception. This means you have to know in advance whether the phenotype is there, whether the genetics are in place.

Unfortunately, mere registration does not greatly enhance this knowledge, and yet phenotype and genetics both are knowable.

Tell by Looking

You can tell by looking more than most writers put in the opening sketch of a novel. You can single out any one animal in any pasture and tell whether she's ever aborted. If the nipple is wider at the tail where it comes out of the udder — she's aborted.

Most of the time she'll be that way in all four nipples. You can walk into any herd of heifers and know whether she's cycling or not. When a heifer is nursing, her mother is putting a lot of oxytocin into her, she has a full udder. The oxytocin she receives from her mother is causing her to fill up. As soon as you take her off her mother, she'll lose the floor of that udder. Those nipples fold up like a piece of dead skin. When that heifer starts producing estrogen (ovaries), she'll develop another floor. The nipples become usable, being prepared for work.

At the point of conception, also, the hypothalamus is determined, as is the pituitary. The pituitary controls the adrenal gland.

The presentation of the adrenal is indicated by a swirl of hair. You can preg check by that pancreas swirl.

The liver stores nutrition. Basic metal nutrients — copper, zinc, selenium, magnesium, all have to be processed and put in the liver, stored and excreted, governed by a process called homeostasis.

Next is the thyroid.

The pituitary causes the thyroid to function. If the pituitary is undeveloped, then nothing works properly. Late development of the pituitary, missed pregnancies, late maturing animals, and settling, all terminate herd profitability.

It is located under the neck back of the joint. I can look at the eyes of a cow that has retained placenta and see that fact. Her eyes will be sunken; it'll take a year for these eyes to straighten out. If she's been ill, her eyes will sink. This means the thyroid is not functioning properly.

Long hair on the neck indicates a thyroid problem. That animal is not secreting the hormones and enzymes required to make an animal fertile. This puts a fertility problem into the equation.

The pancreas gland has its sign. It is located on the side, straight up from the navel.

I have never seen a cow or bull that did not exhibit this pancreas swirl. As she produces progesterone upon entering into pregnancy, the growth on the side becomes further evident and very noticeable. You can literally preg check a cow from the vantage point of a pickup truck.

Pancreas swirl on bottom of belly in these Murray Grey cattle.

A pancreas gland exhibited by way of that sign on the side tells anyone feeding cows whether there will be an abundance or paucity of eggs. I have seen the count hit ten or more. When the pancreas sign is on the bottom, on the flat part of the stomach, the egg count will be five or six.

These Observations

As I relate these types to the reader, I am startled by the necessity of stepping forward, reaching back, then jumping over the landscape as if to reiterate that everything is related to everything else.

I've worked with a lot of problem cows over the years to try to get them pregnant. I developed the ability to do that. I have to tell you that I did it with minerals. When someone brought me a cow with a round ovary — round and flat — there was no recovery. I have ordered out a lot of blood work through the agency of Dr. Robert Coffee. In 100 percent of the flat ovary cows there was a magnesium deficiency. This link to pregnancy has still to be established scientifically. I suggest it will be established.

A late maturing pituitary gland merely tells you all other glands will develop late. The animal in question simply will not work for you.

The secret is *did she cycle when she was eight, nine, ten months of age?*

If the hair is standing up in the adrenal swirl, the animal is cycling. I need to digress only for a moment. The baby calf is born with flat shoulders. By the time it is three or four months of age, a change becomes evident — if it is ever to happen.

Morphology

Morphology is determined by complete interaction between heredity and environment. Externally affected organs, skeletal size, muscling, fat deposits, hair coat, color and breed, all are characterized by the morphology of the animal.

The failure of an animal to shed telegraphs problems to come. It indicates a gland system not functioning properly. The fault can be genetic or management. No animal is genetically disposed to withstand starvation.

I can confirm a pregnancy rate increase from 65 to 80 and 85 percent merely by adding 14 mg selenium to a good bag of minerals.

This done, cows shed, at least in my Arkansas area. Later on I will detail why when selenium is low, copper, zinc and manganese cannot work properly. Animals that won't shed are screaming for attention.

Needless to say, problems vary by area, always with indicator signs asking for attention. Grass is the answer if it carries a payload, otherwise alternative mineral supplementation can be used to fine-tune the ration. The cow is a veritable catalog of indicators. Right beside the anal orifice, there is a bald place with a black spot that seems to invite flies.

The presence of flies tells you that the sebaceous fluid is out of balance. There is a gland at that location, and flies seem to love that sick sebaceous fluid.

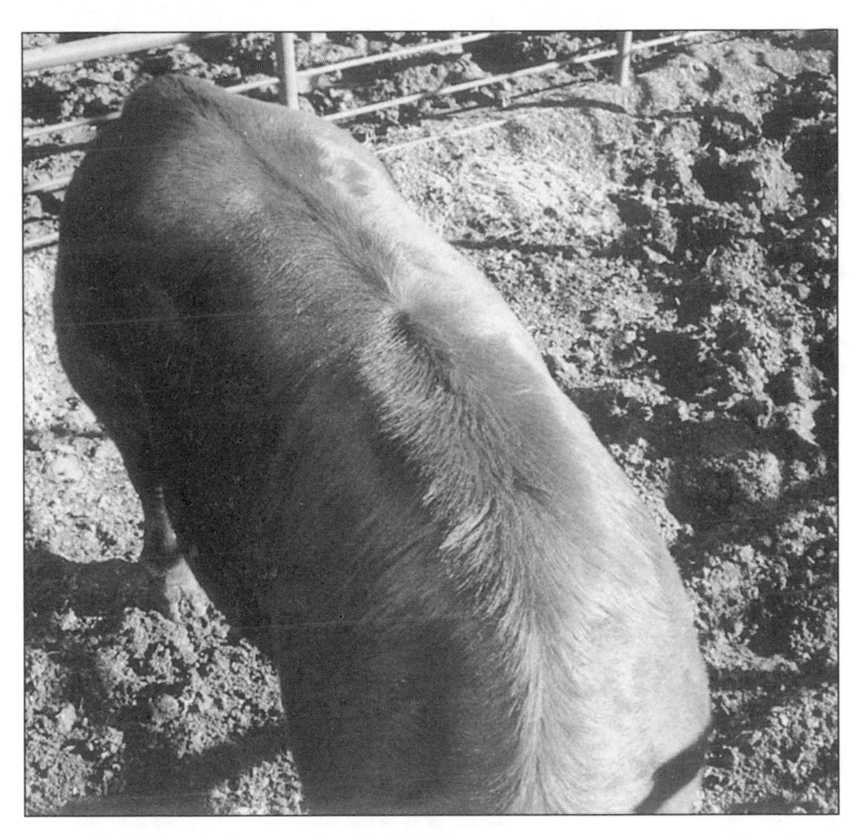

An adrenal swirl atop the back. This area is sensitive to phosphates used to control grubs.

But when the liver is working properly, the pituitary and gland system erase that dark spot. When the sebaceous fluid is healthy it repels flies. The great Professor Phil Callahan says a wave in the infrared portion of the spectrum invites or repels insects, which is why balanced health repels and sub-clinical illness in an animal invites the several types of insects, each with its own calling signal and DNA.

Fly Control

This tells us that fly control is seated in pasture management, mineral supplementation when necessary, and that sprays and poisons have little place in management.

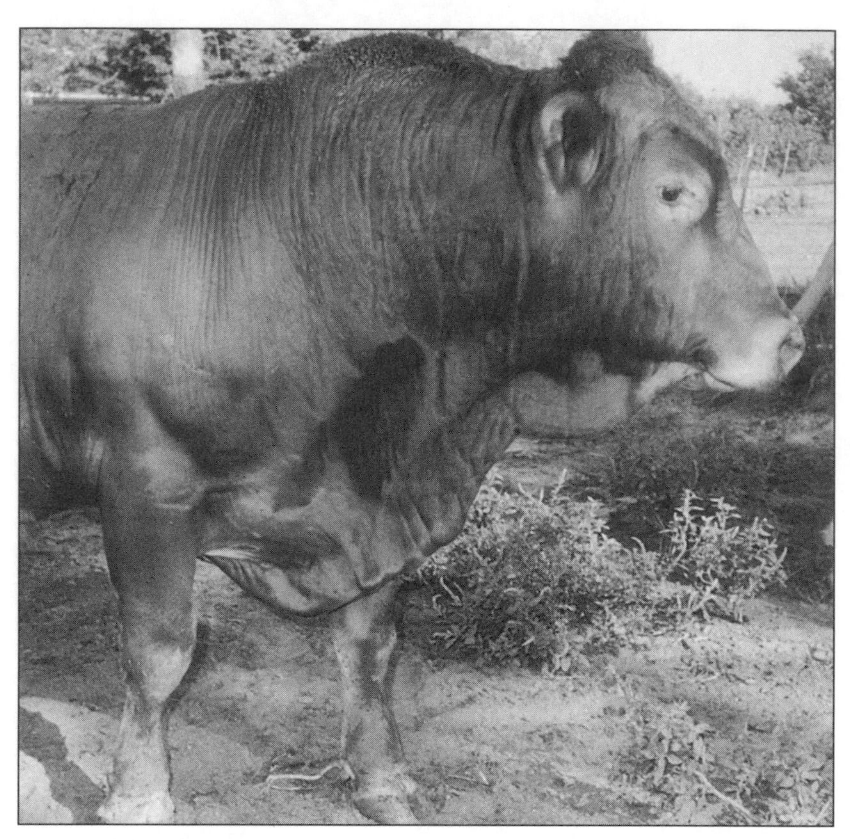

Signs of the thymus.

The pituitary at the base of the brain can be seen in the following picture of a bull with an adrenal swirl.

All the hormones that regulate the ovaries and testicles of the animal are exhibited in these swirls of hair. The adrenal gland is made part of the outward sign system.

The thymus is on each side of the neck. It causes that swirl of hair to mark the side.

There is no outward indicator for thyroid function except long hair and sunken eyes. Otherwise, testing is necessary.

In each of these sections I am prompted to return to retained placentas, mastitis, high somatic cell counts, stillborn calves, weak calves, unthrifty cattle, poor body condition, poor reproductive performance, *E. coli* scours and diarrhea. These things rob the cattlemen of millions of dollars each year. Yet every one of the above problems can be managed out of the herd using pasture and mineral supplementation.

Some 25 Years Ago

Some 25 years ago, I had about 500 head of commercial cows. I would have a minimum of 200 calves with *E. coli* scours. A man with a memory as good as mine does not have to do things twice and chasing calves for treatment was a twice-daily affair. I wanted to know why this was happening, scours being pretty hard to miss. I turned to the books. I found that *E. coli* scours were there because of a low immune system. I discovered that all enzymes have trace mineral keys. Once I learned to read the signs, I knew selenium was too low. My instant remedy was 2cc MU-SE, the result being total elimination of scours.

There is a better answer, of course. Playing veterinarian for new calves would test the patience and endurance of Job. The point is that signs tell you what you need to know well in advance of the traumatic event. Relying on settled information and seeing what you look at are paramount. I can testify that I have never had another case of foot rot, retained placenta, not even a mastitis problem, not for myself or for my clients.

In the chapter on bulls, I will supply more details than many cattlemen can instantly comprehend. For now it is enough to point out that many bulls have more white blood cells than sperm cells in their seminal fluid. Dr. Robert Coffee's analysis revealed that 100 percent of the reports told of selenium and copper deficiency. Using MU-SE to triple the intake of minerals, I could collect semen from such bulls in 50 days. The white blood cells inevitably seemed to evaporate week to week.

It seems that with deficiency, cell walls become thin — especially in the scrotal area — thus the invasion of white blood cells in the seminal fluid. In cows the entry is into the milk duct area. A sort of rotting takes place, causing mastitis. The bull avoids this infection by ejaculating via masturbation each day. While a mastitis-like infection can be avoided, pre-potency is also compromised.

Cause is a most elusive term and concept. Some owners speculate that some of our security measures — five- and seven-way vaccines, hormones, antibiotics — might be adding to the problems of husbandry. Minerals sized too large for assimilation can pump their way through the rivers of the blood system without making it to the hungry cell. The wisdom to link cause to effect is rare. This realization assigns a new importance to the long term and to genetics.

A Bottom Line

In almost any discussion, I try to come to a bottom line. Here it is simply that if a nutritionist must be employed, that professional has to be able to observe the herd and tell you what is wrong. If this person is unable to do that, using either the signs and symptoms discussed above or other methods, an injustice will follow. That is a given.

If that professional exhibits an inability to read those cows, you're in the presence of guesswork. The nutritionist has to be able to tell you what is wrong. You have to see the readout with your eyes.

It may still take lab work to discern the effects in terms of micro-mineralized refinement, but the raw results are as clear as a slide in a projection machine. I have tried inorganic minerals, the kind feed manufacturers import from China and bury in fertilizers and feed rations, always with internal deterioration and semen half-dead, this after a 75-85 percent score earlier in the week. The lesson came through loud and clear. Chelates such as those available through Albion Laboratory, Clearfield, Utah, are highly biologically available.

In the wake of a blunder, it may take a year to stabilize an animal.

Chapter 3

The Mandate to Measure

You have to know the inside of a bull before you can discern the validity of outside observations. I suppose I owe some parts of that observation to Monte Roberts, the rodeo star, horse trainer and author of *The Man Who Listens to Horses.* Roberts was a champion steer roper, bull dogger, bull rider and all-around cowboy before he took to training thoroughbreds for the horse racing track. His wife was a sculptor of the equine form. Artists used a form of linear measurement long before Bonsma and his disciples refined the idea to measure bovine conformation.

Roberts told of watching an artist sketching horses near one of the paddocks. "He was using a triangle to help maintain correct proportions," wrote Roberts. "Imagine a horse seen from the side; and over that horse lay a triangle, its apex sitting as high as a horse's head, and midway between head and tail. From that high point one line angles down at 45 degrees through the shoulder. Another angles down at 45 degrees through the hip. And the base line runs horizontally through both knees and hock."

The objective is to achieve balance with the apex exactly in the middle. The baseline exceeds the length of the equilateral triangle in the thoroughbred. The sculptor's scheme comes to mind when the measure of the bull is to be considered. I have refined my general presentation several times and now rely on what journalists call standing copy, which was presented to the readership of *Acres U.S.A.* in January 2000.

Sizing Up the Herd — The Mandate to Measure

It is common knowledge that the amount of hormones produced by a bull (testosterone) and the female (estrogen) determines the amount of meat a carcass has on it at a year of age hanging or live. The size of the testes determines that for the bull, and the ovary determines it for the female. For either to achieve that maturity level, you must choose and mate animals with that genetic ability. When you choose and breed for desired traits for two or three generations of matings, those traits become homozygous, and the traits you have chosen to achieve are the muscle patterns needed for carcass and a profit.

The procedures for linear measuring were developed and researched in Montana and Nebraska more than 20 years ago. The purpose for the research was to develop a standard for reproductive performance and maintenance efficiency and bring to the surface structural defects, packaged with a quality carcass that is acceptable and profitable. Today we can measure young cattle at 10 to 14 months of age and know with accuracy how they will perform as seed stock or in the feedlot. Failure to choose for reproductive performance and maintenance efficiency will lead to traits being selected via single-trait selection, and that leads to low performance in production areas. These two traits must be in place for other selected traits to work with accuracy and for the animal to stay in balance.

When one uses linear measuring to build a herd of females, they will have calved at two years of age and 95 percent will breed back to have a calf at three years. When you have selected the proper traits in the sire and matched the dam's weakness with his strengths, then 80 percent or more of the progeny will qualify to go back in the herd as building blocks for the next generation.

By the third generation, you have built a herd that is equaled by only a few and is not only acceptable but is in demand. You will be producing bulls that will sire both seed stock sons and daughters. You have bred into those cows and bulls the ability to be predictable and repeatable with performance.

Linear measuring of yearling bulls and heifers has been in use for over 20 years since its development. Its most important function is to help choose accurately for reproductive efficiency and maintenance efficiency, resulting in carcass production. Man chooses the best of his animals for replacement or sale with the eye. The ones he chooses are the most phenotypical as far as what the eye has been trained to see. However, since choosing by eye is very subjective, sometimes the wrong animals are kept in the herd. Linear measurement is a tool, which allows the producer to make a more objective decision that the eye alone can't make.

Cattle that win in the show ring normally will not work in the pasture. They tend to fall apart when they are put out on grass. Have you ever wondered why? It is because these animals have not been chosen for the reproductive and maintenance efficiency, they usually are hard keepers and cannot perform on grass alone.

Linear measurement of your males and females can help you select so that you can produce an animal that is efficient and predictable. There is a specific pattern the male and female should fit for best performance.

Heart Girth — A Growth Indicator

Large heart girth gives increased vigor, increases adaptability, increases feed efficiency and ease of keeping. Large girth creates more space for large heart and lungs, red meat and a larger loin, and it also supports reproductive efficiency.

Insufficient heart girth is a structural defect that allows front feet to toe out. It will make the animal more susceptible to stress and will reduce reproductive and maintenance efficiency. Small girth means small heart and lungs.

For males, an animal's total top line should be equal to or 0 in relation to the heart girth at 10 to 14 months of age. The closer the top line and heart girth, the more efficient, adaptable and vigorous the animal is.

A female's total top line should be even to the heart girth at one year of age. The closer the top line and heart girth, the more efficient and adaptable the animal is.

Rear Flank — A Fertility and Maternal Trait

The flank in the male should be equal to the heart girth or two inches greater. The difference between rear flank and heart girth is correlated to fertility. Rear flank should be at least two inches larger than girth, or larger. A deep flank is a high indicator of milk production, high maternal traits, and will produce maternal daughters. A small difference between girth and flank suggests that you should check other fertility indicators. The flank can vary two to three inches by fill (feed water). A high-flanked cow represents a deficiency of red meat on hindquarters and a higher maintenance cow. High flanked cows are normally more high-strung and low in maternal traits.

Rump Length in Females — Fertility Indicator

Rump length significance is the percentage of rump length compared to the two-thirds top line length. Desirable rump length in the female is 36 to 40 percent of the two-thirds top line or less. With an ideal somewhere at 36 to 40 percent, you maintain reproductive efficiency with maximum meat on the rump. A lower rump percentage, such as 36 percent, indicates a female higher in fertility than one at 42 percent. There is a direct transfer of rump of female to neck and shoulders of her sons. If the rump is too long or short, then the animal will be out of balance. If percentage is over 40 percent and the cows become coarse, attention should be given to size extremes.

Rump Length Percent in Males

Rump length percent in the male influences neck length of his daughters. Rump length has to be in balance with masculinity. Preferred is 36 to 40 percent for proper balance; 38 to 40 percent is ideal. Less than 36 percent in the male increases the neck length of his daughters. Less than 36 percent creates a loss of red meat. Less than 36 percent creates a lower standard for masculinity and femininity. Over 40 percent causes his daughters to be too coarse in the shoulders, a loss of femininity and a loss of maternal traits. There is a direct transfer of shoulder area of the male to his daughter's rump. She will produce masculine sons. Both will have more meat on the frame.

Top Line

The total length of the animal from back of rump to front of poll is the top line (neck length + body length = total top line). The top line is important for balance. The girth of the animal needs to match the top line. Cattle with extra-long body lengths are normally high-maintenance animals. The top line and girth cannot match. The girth will be "pinched" (a concave chest) behind the front legs.

Body length should be two-thirds of the top line. The two-thirds is the length of body and rump from back of rump to middle dip in vertebra between the shoulder blades; the rest is the neck length. If the back is too long, it will make the rump length percentage too short. It will also make the top line longer than the heart girth, causing the animal to be out of balance and suggest a high-maintenance animal (male and female).

Female Neck Length

The neck length should equal half of the two-thirds. The neck, back and rump length should each be the same length. Every inch that the neck length exceeds the other two measurements is an indicator of lower feed efficiency, loss of weight while in milk production, hard doing in cool weather, slowness to breed and late calving. The long neck is a result of single-trait selection. If the neck is longer, reproduction of the cow suffers. The long neck directly affects reproduction and maintenance of the animal. Additionally, the shoulders will be too narrow, and the front feet will be too close where they exit the chest cavity. There will be structural defects (twisted feet, crooked legs, loss of meat and milk production, and problems with reproduction and performance).

Male Neck Length

The neck length of the male should be a minimum of two inches shorter than half of the body length. A neck that is greater than this indicates an absence of male hormones, resulting in late-maturing offspring and it is a good sign of late breeders and the potential to skip calves, lowering the reproductive cycle. The shorter the neck, the higher the male hormones. The higher the male hormones, the more crest development. The cervical vertebrate has a tendency to curve up, making the neck short. As the neck gets shorter, his offspring's rump gets wider, resulting in more pounds of beef and higher reproductive performance. A short neck is a good indicator for libido and higher fertility. A short neck and wide shoulders is a good indicator for uniform gestation lengths and uniform birth weights. As the neck gets shorter, the female offspring's rump gets wider, resulting in early maturity and/or a heavier carcass finishing early. Bulls with a disparity of less than two inches (and, thus, having longer necks) lack the production of male hormones, resulting in late-maturing offspring, indicating slow breeders and late finishing in the pasture and feedlot. A bull with an absence of male hormones never produces offspring suitable for herd bulls.

Male Rump Width

Rump width should be a minimum of 44 percent of rump height. If rump is less than 44 percent, there is an absence of red meat. Wide rumps are an indicator of early maturity and ease of fleshing. The wider the rump, the larger the rib eye and the larger the loin eye. The rump and loin of a carcass has 88 percent of the high-dollar cuts. A bull with less than 44 percent rump width-to-height ratio has an absence of red meat and is normally a higher maintenance animal. The rump area should balance with the neck and shoulders.

Female Rump Width

The rump width should be a minimum of 40 percent of rump height. Femininity is in the rump of the female. The wider the rump, the greater the reproductive efficiency and the earlier the maturity and ease of fleshing. Less than 40 percent will cause late maturity, slowness to calve, and a tendency to skip calves from time to time. Less than 40 percent also demonstrates an absence of red meat (lighter carcass). It will cause her sons to be feminine, narrow or light in shoulders and neck, and have structural defects.

On a more technical scale, the rump width should be at least two-and-one-half inches wider than the rump length. The rump width is the most prominent indicator in the female. A wider rump than its length is the highest indicator of fertility and femininity. The wide rump will shorten the neck length of her sons. The narrow-rump cow has an absence of red meat.

Female Shoulder Width

The shoulders in the female should be the same width as the rump length. A deviation of one-half inch is acceptable. Shoulders that match the rump length represent a very feminine, very fertile female. This also suggests an easy maintenance female and easy fleshing. Shoulders that match the rump will have a larger loin area. Cows with shoulders that match the rump length will produce heavier carcasses. Too narrow a shoulder is a structural defect and denotes a high-maintenance animal. Too wide a shoulder causes a lack of milk production. Too wide a shoulder is a form of masculinity and therefore a loss of reproductive performance, plus a comparable decrease in the amount of red meat in the rump area.

Male Shoulder Width

The shoulders and neck are the most prominent indicators in the male. Shoulder width should be two inches wider than the rump length at 12 months. Wider is better. Wide shoulders are a high indicator of reproductive efficiency, creating a higher live sperm count. Wide shoulders cause early puberty and result in the production of female offspring that calve early, breed back and wean a heavy calf and/or give more milk. Wide shoulders provide the ability to withstand stress. A proper bone structure — not meat, condition or fat, but genetics — causes wide shoulders and short neck. Yearling bulls with a good adjusted shoulder width correlate with uniform gestation lengths, uniform birth weights, calving ease, and more uniform weaning weights, resulting in more pounds of beef or milk.

With narrow shoulders, little or no crest at 12 months, and a feminine look, you lose all of the above traits and performance. The bull controls the birth weight, and it will be within two to three pounds of his birth weight if he was masculine at 12 months.

Rump Height

The rump height is highly correlated to gain ability. The tall, long-legged animal will gain over a longer period of time, will reach puberty late in life, and will be slow to finish, with only a small percent grading choice. The same holds true with male and female. Height destroys reproduction. Height represents single-trait selection. Tall animals tend to be out of balance.

Thurl

The thurl and its form determine the calving ease of the female. Straight hocks are most undesirable and influence the hindquarter form and efficiency. In the straight-hocked animal, the thurls are pushed upward, and this results in the square rump. It causes difficulty at calving, but holds no effect on the amount of muscle or red meat flesh on the hindquarter.

Hide and Hair

The hide of the well-balanced bull and cow is much thicker than in the out-of-balance animal. The thick hide has more vascular flow under the hide, resulting in shorter, thicker hair and more insect repellence than the animal with thin hide and dull hair. The balanced animal with a thick hide will have a well-developed panniculus muscle and a sensitive pilomotor nervous system that will move at the slightest irritation by insect. The animal whose hair is short and thick and stands straight up will have less insect problems. The animal with a thick hide stays warmer in winter and cooler in summer. The hide forms the barrier between the animal and the environment, ultimately determining the adaptability of the animal. The hair coat accurately reflects the well being of the animal, its hormonal and nutritional balance. Thick movable hide with short hair secretes sebum that acts as a tick and insect repellent. Early-shedding heifers reach puberty long before those that shed late. Early shedding is correlated to early ovarian and testicular activity. Animals susceptible to external parasites are also more susceptible to internal parasites. Less adaptable animals in an environment will normally have higher external and internal parasites. Well-adapted adult animals seldom if ever have internal parasites that are a problem, and deworming is not necessary.

Udder and Teats

Functional efficiency and soundness of the udder, which is a critical factor in the productive lifespan of cows, is critically dependent on formation features such as pendulousness, size, shape, placement, teat shape, teat opening, and pigmentation of the udder. The scrota and testicular conformations of bulls is responsible for quality of udder on daughters.

Pigmentation protects against cracks (chapping), painful milking and sunburn. Too large an opening predisposes the animal to mastitis. Too small an opening causes bruising during milking. The size and shape of the teats accurately reflect the hormonal balance and activity of the cow and heifer. Cyclic and ovulation inefficiency increases the size of the base of the teat and udder development in the non-pregnant heifer and is a sign of cystic ovaries. There is little udder development in a heifer that has not come into heat yet. The teats are very small and pushed back into the skin and folds of the udder. After the heifer cycles a couple times, the teats hang down perpendicularly from a fairly well developed, firm maiden udder. Heifers that have aborted no longer have the maiden udder. The udder is much larger; the teats are well developed and are typical of teats that have not been suckled. After the estrus cycle starts, the udder should have shed all of the long hair, which should not return again. The udders are broad at the base and tapered to a fairly sharp point. The older cow that has aborted will have a yellowish creamy plug in the teat. The cow that suckles a calf has a well-formed teat, which is very smooth as the butterfat is massaged into the teat when the last milk is sucked in the sucking process.

Testes and Scrotal Sac

The testes of the bull should be 38 to 40 centimeters or larger at 12 months of age. However, 38 centimeters does not always ensure a bull will pass a fertility test. Bulls with 38 to 40 centimeter testes normally won't get any larger than 44 centimeters. A matured bull with large testes (44 and up) normally is a late-maturing bull. Large testes are easier to bruise. You should never use a bull with one testicle. Hypoplastic (one testis) bulls normally have the conformation of a steer in hindquarters. This is because of prolonged growth as a result of low hormone production. Pigmentation of the scrotum makes it resistant to warts, sunburn, ringworms and infections. The scrotal sack should have a buckskin look, with only a small amount of fine short hair or no hair covering. Nipples should never appear on the scrotal sack neck.

Bulls require a high degree of physical fitness, which is never attained with obesity. A masculine bull is easy to keep in shape. The heat exchange system in the spermatic cord becomes ineffective with the presence of fatty tissue, resulting in a reduction in sperm motility and an increase in abnormal spermatozoa. The testes should be well formed (both alike) and hang straight down and not tucked up in the back. The scrotal sac should feel a few degrees cooler than body temperature for normal spermatogenesis to take place.

Notes of Interest

A cow should deliver a calf that is seven to eight percent of her body weight.

- Some heifers will give birth to calves up to nine percent of their body weight.
- A cow should wean 55 percent of her body weight or more.
- Calves must have an umbilical cord at birth.
- Heifers should be cycling at 10 months of age.
- Bulls should have at least 38-cm testes at 12 months.
- Bulls should be able to breed and settle 80 percent of the cows serviced on first service at 12 months of age and not lose weight.
- Bulls that are masculine at 12 months should sire calves the same as their own birth weight.
- Heifers should conceive at 14 months of age.
- A bull or cow that does not shed long hair cannot achieve the highest potential.
- Bulls and cows should have pigmented skin and dark hooves.
- Hide and hair are indicators of adaptability.
- All animals must have a thick, movable hide.

- Short, thick hair is more fly and tick repellent; dull, long hair attracts flies and ticks.
- A highly fertile bull will have darker hair on the lower half of the body.
- The bull and cow should have a wide mouth with large nostrils.
- The cow and bull should have pigment around the eyes.
- The cow and bull must have a large gut capacity.
- Cows and bulls that can survive on grass and hay and which breed and calve regularly are economically valuable.
- Genetics are set in place at the point of conception — the breeder must know his genetics.
- When you create a large bovine body, the gland system does not get larger.
- The ideal size of cows is 1,000 to 1,100 pounds.
- Selecting for birth weight, milk, height, length, weaning weight, or any one trait is considered single-trait selection.
- Selecting for reproduction and maintenance creates all good necessary traits.

At the point of conception all things are determined — reproduction, maintenance, udder, bone, frame, feet, legs, ability to pass traits, hair, color and temperament. What follows is a review of the above for each sex.

Guidelines and Correlations — Female

Top Line

The top line is the total length of the animal from front of poll to back of rump and is taken by three measurements: neck length, body length, and rump length. These three measurements equal the total top line length.

Linear Measurement
FEMALE

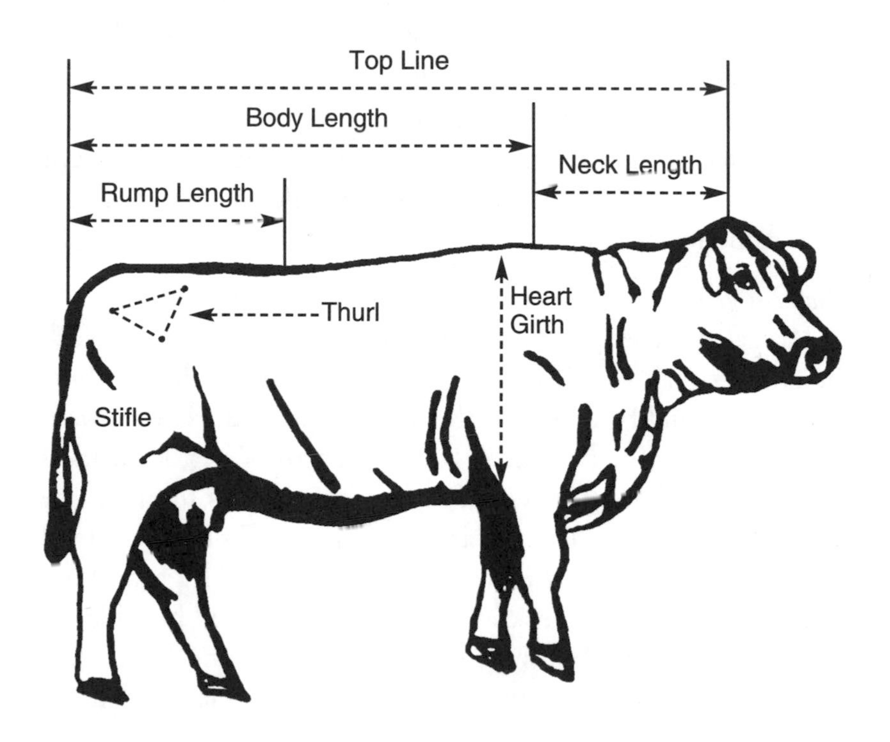

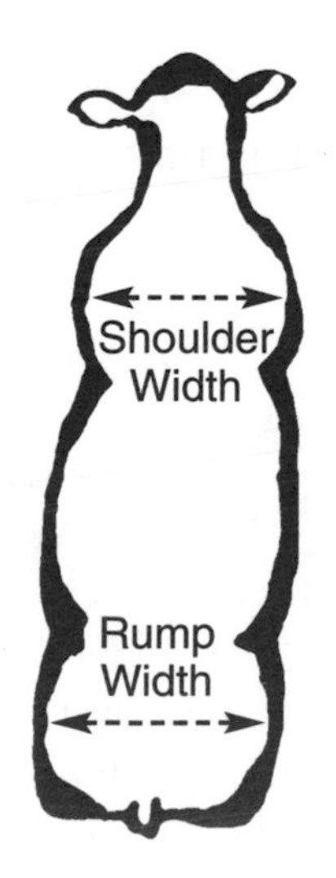

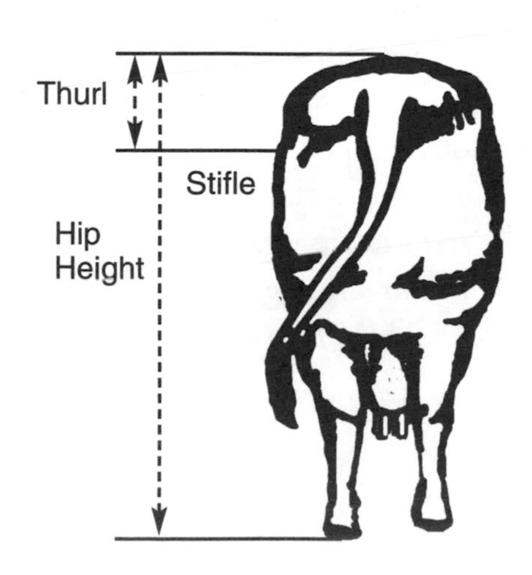

Heart Girth

The total distance around the animal's heart girth should be equal to the total top line or larger at 12 months of age. Large girth is needed for proper size of vital organs (heart, lungs, glands). The closer the heart girth is to the top line, the more efficient, adaptable and vigorous the animal will be. Insufficient heart girth is a high indicator of structural defects, allows front feet to toe out, hooked toe, more susceptible to stress and is a high maintenance animal. Reproduction suffers.

Neck Length

Actual neck length minus one-half the body length. Neck should be equal to half the body or two-thirds of the top line. The neck should be one-third the total length of the cow. A well-balanced cow will have a neck half the length of her body. Ideal range is plus or minus one-half inch. If the neck is too long the cow will be high maintenance and easy to stress. Long necks tend to over-produce milk, making the maintenance high and are slower breeders. If the neck is too short the cow will be wider in the shoulders (coarse), lower in milk production and have a loss of femininity. Long necks in the female are not symbolic of femininity.

Body Length or Two-Thirds Top Line

The two-thirds top line is composed of the rump and back length or the distance from the middle dip in vertebrate between the shoulder blades to back of rump. If the back is too long it will affect the neck length and the animal is out of balance. Long backs tend to be weak and will sway. Most long backs have too small a loin muscle. The animal will break behind the top of shoulder. This break or dip is a structural defect.

Shoulder Width

Shoulder width should be the same as rump length. Ideal range is plus or minus one-half inch. Too wide a shoulder in a female will cause a lack in milk production. Too narrow a shoulder in the female requires more maintenance and results in reproductive problems. In the cow, the shoulders should balance with the rump.

Rump Length Percent

This is the percentage the rump makes up of the body length for the rump length percent. In females the rump length should not exceed 40 percent of the two-thirds top line. Thirty-eight to 40 percent is the ideal range. Either side of this is in the extreme. The rump length sets the standard for femininity.

Flank Circumference

This is a fertility and maternal trait indicator. The larger the flank circumference is compared to the heart girth, the higher the fertility. High-flanked cows have a tendency to be a little more flighty, have less meat on the rump, and have a tendency to be higher maintenance. The flank area should be two to 12 inches larger than girth at 12 months. Larger is better. A small difference or less suggests that other fertility indicators should be checked. Fill can affect as much as three inches.

Rump Width Percent

Rump width divided by the rump height (21-inch rump width divided by 48-inch rump height = 43 percent rump width). Rump width percent indicates ability of self-fleshing, ease of keeping, reproductive efficiency and is an indicator of volume of flesh on rump. The minimum rump width percent is 40 percent of rump height. The wider and deeper the rump and flank the higher the maternal characteristics. The wide deep rump represents femininity and reproductive efficiency and these cows have sons and daughters with wide rumps.

Adjusted Rump Width

Rump width minus rump length. The rump width should be 2.5 inches wider than length. The wider the rump is compared to length is a high fertility indicator. The wide-rumped cow has sons with short necks, daughters with wide rumps, has more meat in the rump area and produces sons and daughters with more meat. The progeny will mature earlier and finish by 16 to 18 months of age on grass.

Rump Height

Rump height correlates highly with gainability. Tall animals tend to be out of balance, slow to come into puberty, thus lower in fertility and reproductive efficiency. For overall performance and finishing on grass, a frame score of 2.5 to 4.5 works best.

Thurl

This measurement is taken from ground to stifle joint to top of back. Thurl should be 13 percent of the rump height or greater. Greater is better if the slope of the rump is of the proper angle. It indicates pelvic depth and structural soundness of hind legs. If the thurl is properly in place the animal will track (back foot in front track). A properly structured thurl makes for ease of calving.

Udder

The udder should be small and tucked neatly between the back legs with four equally placed nipples, each three to four inches long. The udder should attach high up behind the back legs for longevity and soundness, should blend into the lower part of the belly very smoothly with no V or crevice between the udder and stomach and should not be tilted up in front. Tilted udders are a structural defect resulting from the sire and his scrotal makeup and have less milk. The mother's udder has a direct influence on the scrotal makeup of her sons.

Guidelines and Correlations — Male

Top Line

The top line is the total length of the animal from front of poll to back of rump. The top line is taken by three measurements — neck length, body length, and rump length — and these three measurements make up the total top line length. The true, total top line for bulls only = 2/3 top line x 1.5 (.5, 1.0,

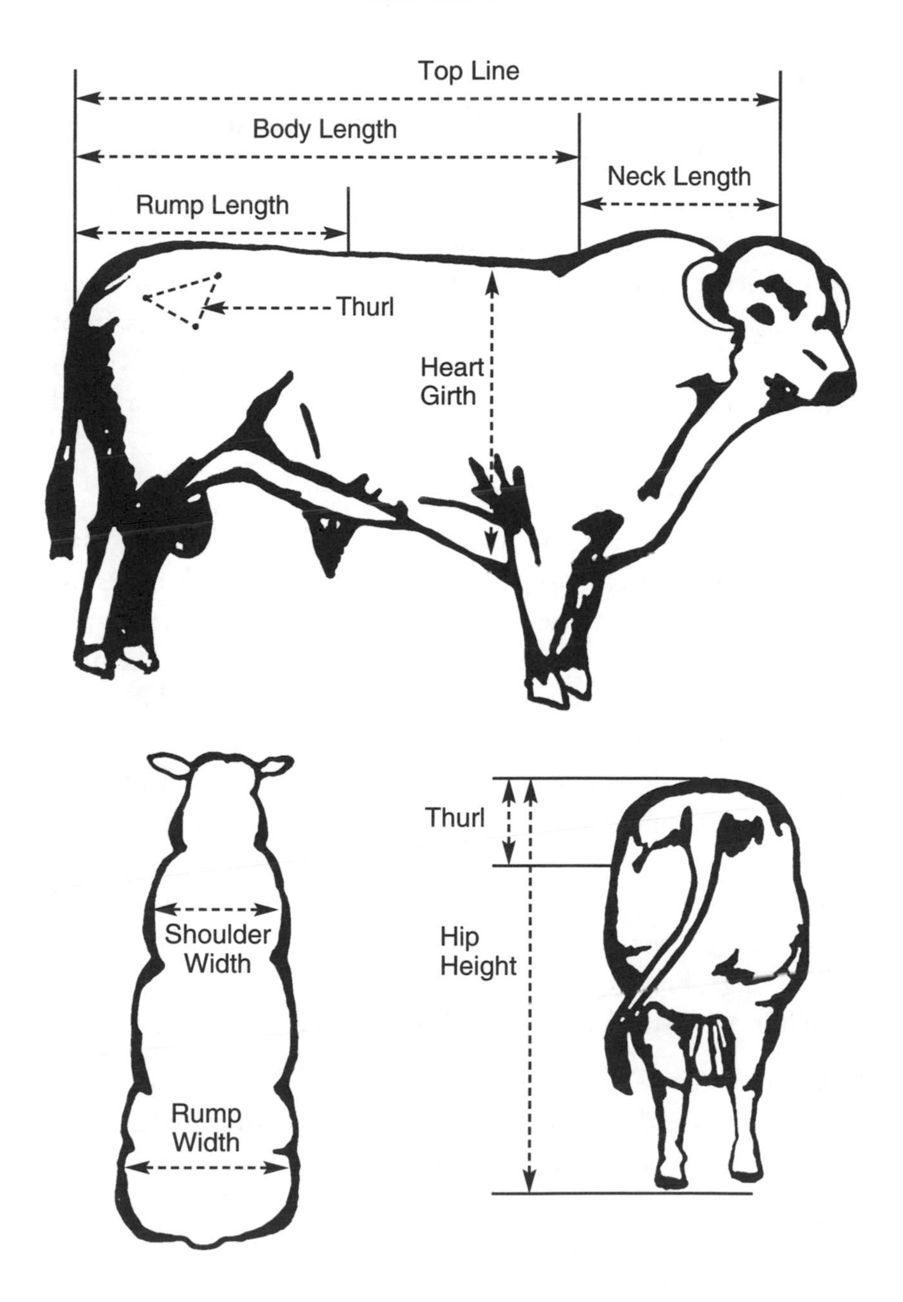
Linear Measurement
MALE
Top Line
Body Length
Neck Length
Rump Length
Thurl
Heart
Girth
Shoulder
Width
Rump
Width
Thurl
Hip
Height

1.5). For example, if the bull was 70 long and his 2/3 is 48, his neck length would be 22 inches long (70 inches – 48 inches = 22 inches). The formula used multiplies by 1.5 (1.5 x 48 = 72) you get the true top.

Heart Girth

The total distance around the animal's heart girth. The heart girth should be equal to the total top line or larger at 12 months of age. Large girth is needed for proper size of vital organs (heart, lungs, glands). The closer the heart girth is to the top line, the more efficient, adaptable and vigorous the animal. Insufficient heart girth is a high indicator of structural defects, allows front feet to toe out, hooked toe, animal is more susceptible to stress and is high maintenance. They do not perform well on grass and reproduction suffers.

Adjusted Neck Length

Adjusted neck length equals actual neck length minus half the body length, *i.e.,* 24 – 22 = 2. The higher the levels of testosterone, the more crest development, the shorter the neck. The cervical vertebrates have a tendency to curve up, making the neck shorter. Short necks are a good indicator of libido — they indicate a larger scrotal area and therefore a higher level of hormones and indicate wider shoulders (male characteristics). Bulls with a neck longer than the 2-inch factor at 12 months lack male hormones, resulting in late maturing daughters, reach puberty later in life and tend to be slow breeders. Long necks are structural defects, no genetic progress can be made with these bulls. They will be high maintenance bulls. Bulls with shorter necks sire daughters with wide rumps and daughters that mature early.

Body Length or Two-Thirds Line

The two-thirds top line is composed of the rump length and back length or distance from the middle dip in vertebrate between the shoulder blades to back of rump. If the back is too long it affects the neck length and the animal is out of balance. Long backs tend to be weak and will sway. Most long backs have too small a loin muscle. These longer animals will break behind the top of the shoulder. This break or dip is a structural defect.

Rump Length Percent

The rump length percent is the percentage the rump makes up of the body length or the two-thirds body length. Divide the two-thirds body length into the rump length. Rump length percent in the bull influences the neck length in his daughters. Less than 40 percent rump length in the bull increases the neck length of his daughters and makes for a smaller heart girth, thus higher maintenance. A range of 38 to 40 percent is ideal, less than or more than is an extreme. This range works best and will make good milking daughters. The rump length percent sets the standard for the degree of masculinity in the bull. If the rump length is below the 38 percent level, the standard for masculinity is much lower. The low standard does not create rugged bulls and in most cases the scrotal area is less than 38 centimeters at 12 months.

Rump Width

Divide the rump height into the rump width and this equals rump width percent. This number should be 44 percent or greater. Greater is better. (Example: 24 rump width divided by 50 rump height = 48 percent). High rump width percent is indicative of early maturity and ease of keeping. Bulls with higher rump width percent usually have wide shoulders, deep chest and a more acceptable scrota.

Adjusted Shoulder Width

Shoulder width minus rump length equals adjusted shoulder width, which in bulls should be 2 inches or greater at 12 months of age. Greater is better. Wide shoulders make room for vital organs. Masculinity is responsible for gestation length, and is the highest indicator of reproductive efficiency and ability to withstand stress. These bulls sire sons when mated to proper phenotype cows (feminine) that make herd sires and daughters that reach puberty early, breed back and wean a heavy calf. Yearling bulls with good adjusted shoulder widths correlate with uniform gestation length, uniform birth weights, ease of calving and more uniform weaning weights. Wide shoulders usually mean a larger scrotum. Masculinity in the bull is in the wide, deep shoulders and short neck with a rump to match the front. Masculinity sires more pounds of beef. In the bull the rump should balance with the shoulders.

Rump Height

This correlates highly with gainability. Extremely tall animals tend to be out of balance due to slow puberty development which lowers reproductive efficiency. Taller animals have less meat on their carcasses than the shorter, thicker animals. Taller animals will tend to have a smaller chest and shoulders and are easy to stress and are a higher maintenance animal. Taller animals do not do as well on grass and take longer to finish. They require more high energy. A frame score from 4 to 5.5 tends to work best for finishing on grass.

Flank Circumference

The flank circumference should be at least two inches larger than heart girth. Small flank measurement is indicative of absence of meat on rump. The same will be passed on to the progeny.

Thurl

Thurl should be 13 percent of the rump height or greater. Greater is better if the slope of the rump is of the proper angle. Indicates pelvic depth and structural soundness of hind legs. If the thurl is properly in place the animal will track (back foot in front track). Measurement is taken from ground to stifle joint to top of back.

Scrota

The scrotum of yearling bulls should be 38 to 40 centimeters. *Larger or smaller is in the extreme and should not be tolerated.* The scrotum should be football shaped with the epididymis on the very bottom of the testicle. If the epididymis is anywhere else but on bottom, the result shows up in the quality of udder of the daughters. The sons will have the same problem as the sire or worse (recessive genes). If the nipples of the sire are on the neck of the scrotal sack, the daughters' udders will be tilted up in front and lower milk production. If the nipples of the sire are on the scrotal neck, his sons will have a large navel area with the opening pointing toward the ground and their fertility is lower.

The four most important areas of the male are shoulder width, neck length, scrotal size and conformation, and heart girth to top line.

Chapter 4

Bull Fertility

As I study the history of the cattle industry, I try to get in the heads of our forefathers and older men living today through conversation and their writings. Those great men left for us guardrails against the quality and type of red meat protein we are producing today.

Our forefathers did not need DNA markers, cloning, or any other scientific research devices to know when they had a superior genetic line of cattle that would produce on grass and hay. They had animals and bloodlines that provided a healthful and delightfully good eating experience on grass. They had no grain, nor was it necessary, to feed 120 days to create fat. History records they had the guts to eliminate a bloodline that was not a superior producer. It seems as though reality and wisdom have not been a part of the equation in cattle breeding over the past 30 to 40 years. It also appears as though we have entered into a silent era in agriculture — and especially the livestock industry — for trying not to meet the demands of the consumer and therefore not making a profit. In reality, the seed stock industry (bull providers) is not in tune with the needs of the commercial producer or the consumer.

A Premise

As I work with the commercial producers across the country I hear the same words coming from all of them. "Why are my calves not consistent? Why do I have calving problems? Why do I have breeding problems (open cows)? Why are the heifers I maintain for herd replacement continuously inferior? Why do I have sickness that robs my profits? Where can I find bulls that meet and exceed my needs? Why is the meat and other food at the grocery level so unacceptable? Why is it, when I produce calves that do well on grass and finish at 16 – 20 months, the markets do not want them? Where can I sell my calves and be paid for their quality? How can I stay on the ranch or farm and sustain my family?"

The average seed stock producer has been producing bulls for seven years or less. How can anyone build genetics or learn the industry in seven years? If the business world operated the way the cattle industry operates we can only guess what would become of it — probably bankruptcy. I have helped producers purchase bulls from some of the most noted breeders as well as average breeders of seedstock. There is little to no difference in the performance of the animals, the genetics or the calves they produce. Most are inferior to commercial cattle. One thing I consistently do find is that when a little breeding pressure is put on the bulls they fall apart (lose most of their flesh). Some go sterile and leave cows open. If you only produced 70 percent on the job how long would you keep your employment?

The genetics of today only produce cattle that require extra high energy. Most of our cattle cannot finish on grass. They are created for the feedlots. Beef cattle have the phenotype of the dairy breeds. The commercial producer has too many open cows, thanks to seed stock bull providers and the grain industry. The heifers from these bulls rebuild herds that are extremely low in reproductive performance and produce high maintenance cows. The feeding of corn and high energy grains, heat processed feeds (mixed rations, pellets, cubes, cakes) reduces reproduction in the cow and bull.

A bull that cannot leave 85 percent of the cows he breeds with calves in the first 21 days of the breeding season should get a one way ride to the slaughterhouse. Every cow left open for 21 days cost the producer 47 pounds x $1.00 = $47.00. This income is lost forever. Do the math. One hundred cows at 60 percent conception in first 21 days. Forty open cows x 21 days x $1.00 pound = $1,880.00 = 21 days loss. Can you live with that loss? There is a major bull and breeding problem in the cowherds of America today. Seed stock producers are not willing to change and most do not know how to change. So what do we do now?

In every herd of commercial cattle I work with I find that 10 percent of the cows are superior and always produce a better quality calf compared with their herd mates — even with the same bulls being used. Those commercial cows are no more mongrelized than the seed stock producer's cows. In most cases they are as genetically superior because of their breeding practices. The seed stock breeder only has ten percent good cows. He uses heavy grain to cover the visible genetic flaws and then keeps the quality bulls and heifers produced for himself.

Program

Commercial producers can establish small herds with those superior cows for breeding. The bulls needed to mate those cows are not for sale. Though semen is available on most genetically superior bulls, those superior bulls are not found in any semen catalogue. They are on the farms of wise old men who never lost sight of reality. The cows are artificially inseminated with the semen of those bulls.

The results are tenfold:

1. The commercial producers are producing their own bulls.
2. They are not paying high prices for inferior bulls (genetics).

3. The heifers from those superior cows and bulls are better than their mothers.
4. The quality and type of the produce will increase as the breeding program continues.
5. The cows and calves are being sustained on grass and some hay in the cool months.
6. The quality of the meat is tender and tasty and is far superior to any other.
7. The demand for this quality of meat is 1,000 times greater than the supply.
8. You are building genetic concentration into your bull and cowherd.
9. You are building a grass-based cattle herd (changing the industry).
10. There is great satisfaction and contentment with what you are creating.

Some are selling bulls to their neighbors. This practice is catching on and is allowing the commercial producer to have the quality of bulls he needs (genetically). He produces the bulls on grass. This makes the selection process much easier and much cheaper. He also watches the bulls and heifers develop and mature. There is no guess as to what may be covered up in EPD's, auction sales, schemes and untruths. Within six to eight years he will replace his herd with the type of cow that will perform well on grass and produce great meat with no equal in both texture and taste. The producer knows his genetics. He can only guess at what he buys. He has the flexibility to mix and match the different body styles (phenotypes) for genetic concentration in his herd to produce the kind of calves that perform best and consistently on grass and there is no spread in the carcass pounds.

This much stated, let's move on into a consideration of bull fertility.

Open Cows

The major reason for open cows in any pasture is low quality fertility in bulls. In most cases it is quite possible to inspect the bull before you pull semen so you'll know when that sire can settle cows. Dovetailing this knowledge with nature's way on grass requires a different kind of cow. The finest foliage will not excuse the maintenance of sub-standard animals. Without the right animal, production arithmetic will allow only a low return on investment. The return on animals in America's pastures today is well below the return possible with the right phenotype genetics.

James Drayson

James Drayson tested bulls for 35 years at Big Sky Genetics in Billings, Montana. He tracked some 1,500 bulls from birth to death and he made a meticulous record of everything that happened. This included measurements twice a year, routine semen analysis and archival records at every stage and interval. In the process he determined standards for young bulls at age six to seven months. Using his records he concluded that it was possible to foretell what maturity would bring and how potent the semen would be.

An appropriate balance of nutrient was assumed, of course. Normal sperm cells require as much. The absence of proper nutritional support can be blamed in the field both achieved and scientific.

There was a time when most of the people in the cattle business were born and raised around cattle. After World War II a lot of people unloaded their money on land. They bought cattle, but they knew nothing about genetics. They knew nothing about mixing and matching bulls and cows. Also, the exotics and out crossings came on scene for high abnormals, broken tails, low motility — incidental abnormalities that prevent a bull from settling cows.

Nature hasn't changed over the past 50 years, but man's perception of husbandry has. In fact the word husbandry was dropped from college curriculums approximately 50 years ago. Industrialization involved agriculture and the United States Department of Agriculture declared the factory in the field public policy. The exotics were big and different. Their presence supported cross-breeding as never before practiced. Advertising and PR spin seemed to say that anyone not cross-breeding was anti-science, anti-university, and against the laws of scientific advice. The mere suggestion of line breeding was tagged as incestuous, the seed of dwarfism and biological distortions. I think I am well within my mark to say that postmodern folklore has captured the minds of breeders during this half century period. In its effort to erase the wisdom of ages, it instilled an ongoing blunder that threatens to mongrelize the entire cattle herd of America.

I refer to cowman savvy when I tunnel back to the 1930s and 1940s to cite wisdom that has been diluted and squelched by the likes of expected progeny difference (EPD) studies. Few of the wise old cattle breeders are alive today.

To survive in the cattle industry, the primary producer must use bulls that are masculine and potent. The cattle produced must be adapted to their environment with efficiency. It takes the correct genetics to produce a uniform, consistent red meat product that meets or exceeds the consumer's demand.

Commercial Cattlemen

It is a sad thing to say that most commercial cattlemen couldn't care less what kind of product they've been producing. The conventional objective is to achieve gain to get the most pound per day metabolically possible. The problem is that the faster the animal grows, the sorrier the meat.

I'm not hammering gain. Gain is necessary and a legitimate objective. But when you target and then obtain five and six pounds of gain a day, the end product will not be very palatable.

For these reasons, and for reasons still to be stated, the primary problem in cow and calf production is a lack of genetic knowledge. Numbers tell some part of the story.

• Reproduction is more important than price. A one percent increase in the weaning rate is twice as important as a one percent increase in price. This means a one percent more calf crop. For a 100-cow herd, the translation is ten extra calves.

• The reproductive rate is 50 percent of the economic value of the farming operation, this according to an Australian study.

High Maintenance, Low Fertility

These few digressions are set down to furbish and refurbish the proposition that low maintenance and high fertility go hand in hand. The low maintenance cow means low to moderate milk production, high lean body meat, and high body fat mass. This is startling to some people in an age that wants to be done with fat of any kind. Fat is energy. Cows need energy to get pregnant.

The same general observation is true for bulls. The university seems to be locked into a "theory period," in which one objective is getting rid of fat. Shelves are swaybacked with data on how to decrease subcutaneous fat, research paid for by the very people it seeks to undo. Yet the hard truth is that this knowledge will run the produced seed stock to terminal slaughter.

In the movie *The Mask of Zorro* the hero told a young man, "When the people are ready, the master will appear." Such statements are enigmatic, but I think that saying means that the master arrives the minute people do their own thinking.

Want of Ideas

"It was the best of times, it was the worst of times...." Charles Dickens used these lines to open *A Tale of Two Cities,* but I submit they apply equally well to the cattle industry today.

Our best of times has to do with genetics. A dour view of the matter probably would lead us to believe that the point of no return has been reached. I do not agree with this. I think there are some bulls out there and some cows that can inoculate a reversal of the industry's trail. Needed is a standard, not only for the sire, but for the dam as well.

The glandular system, it has been observed, is the cow's motor. If it is not working, then the cow is not working. The linear measurement supports that standard.

Having examined signs and symptoms and having measured according to the Bonsma standard, we now come to a top, not bottom line. The problem with most of the bulls I've examined has been low fertility.

There are five levels for breeding of a bull. These are optional, tolerable, objectionable, undesirable, and unacceptable. Trouble beckons the minute the first two are passed and the inevitable result is that there will be a lot of open cows.

A Young Bull

I like to test young bull calves at seven to nine months; this norm is according to James Drayson. Drayson found, by following bulls through their life cycle, that if a bull calf did not have a scrotal length of five to five-and-one-half inches at seven to nine months of age, the calf was a bull in gender only. That animal will never perform. Other structural defects go along with this observation. A scrotum that does not measure up is in fact a structural defect, and not much different than having a deformed leg or an askew nose or toes that won't track.

Drayson did not pull semen at that age because the animals were not producing semen at that age.

The second step in the Drayson ladder was tolerable. The average tolerable bull is four-and-one-half inches in scrotal length. The length of the testicle is as or more important than the circumference.

That single statement ought to be cast into bronze and nailed over the office door of breeders who hope to guide progeny in terms that are holistic — not single target trait — of any breed. All this has to do with the pituitary. The hypothalamus regulates the pituitary. The pituitary regulates the testicles, ovaries, liver, adrenal glands, pancreas, thymus and the thyroid. Those are the power centers that drive the animal.

At 11 months of age this bull impregnated 21 of 24 heifers in the first 21 days. Owned by Tom Zimmerman.

If the animal is a heifer, she is cycling by the time she is ten months of age, preferably six to eight months of age. As for the bull — can he physically settle cows when he is 12 months of age? If both are in place, the pituitary has accomplished its purpose. If the pituitary has functioned properly, only the size and shape of the scrotum can interdict the correct result.

The Optimal Bull

The optimal bull at 12 to 16 months has testicles between six and seven inches in length. They'll be between 38 and 40 centimeters circumference. That bull at that age is breeding cows. He'll have 980 to 13 million sperm cells per cc. The approximate live count will be between 75 and 90 percent. He'll settle 80 to 90 percent of your cows in the first 21 days of the breeding cycle. Even a sorry cow that cycles will become pregnant with the above-described bull, whatever the breed.

Most statements without proof or demonstration are worthless. For this reason I call as a witness the bull pictured on page 65.

When this bull was eleven-months old, Tom Zimmerman turned him out with 24 heifers. He got 21 of them pregnant in the first 21 days. The other three came back, two of them to be seeded, one failed. The bull gained 200 pounds of weight while performing his duty. That's the kind of bull the farmer needs. It has been my experience that you can wear out a truck or a Cadillac trying to find that kind of bull.

Masculinity in bulls is expressed predominately by the head, neck and shoulders. When such expression is evident, that animal is capable of passing on its genetics. Without that masculinity, transfer will not happen and problems will be its legacy. The well-made bull that is wide across the shoulders and the cow that is very feminine, that is constructed correctly up front, signals the eclipse of problems.

That washed-out bull so deficient in prepotency capability has a depleted liver. It had just enough in it when that bull left the farm to get by. Stressed by transportation, say 1,200 miles, introduced to a strange environment, the end of the bull is only the harvest. Usually they die. Seldom is there a way of saving such an animal.

Had such an animal been endowed with shoulders and depth of front end, such stress could have been endured. Drayson came by this in-depth interest in prepotency while learning the then new art of AI at Armour Company. Armour had a BCAI program for five years on ranches throughout the U.S. and after 1962 he taught artificial insemination and related subjects in 23 states as well as in two Canadian provinces. Collection, processing and freezing semen caused him to find a vacillation between the endocrine glands and the fertility, sub-fertility or sterility of bulls.

The star pupil in his pantheon of bulls was the one characterized as fertile. It had the 75 to 95 percent capability to settle eligible cows during the first 21 days of breeding.

Sub-fertile designates below normal fertility. This was Drayson's 10 to 60 percent bull, the usual five to 10 percent reduction attending AI procedures.

Sterile designates the term for the bull that settled the cows only five percent, or less, of the time.

Those signs and symptoms discussed earlier come into their own when the cowman makes his eyeball appraisal. Hair texture and pattern, horn color, neck veins, tail structure, scrotum condition and dimensions — all furnish clues and proof of prepotency. Sadly, all are ignored if even understood by show ring judges. Few cowmen ever enter the show ring to judge livestock, that role being preserved for college professors, breed association experts and celebrities with no cow savvy. The cowman must take on the role of judge just the same.

In addition to measurements and signs that punctuate almost every paragraph, special emphasis on hair beckon the observing cattlemen. Drayson's observations are so pregnant with insight, they demand inclusion verbatim:

> *"One of the most disturbing failures which one should look for first is bareness of the hair displayed about the head and face of a bull. The hair shafts should be coarse in appearance and to the touch. The existence of coarse hair is a strong indicator the bull can be placed in the fertile category."*

Hair patterns can be read like a map, and the subtle turn of coarse hair to fine hair — on the head, neck, even on the body — generally indicates faltering fertility. As the hair silkens, prepotency drains away, sometimes rapidly, sometimes over many months. Coarse and wavy hair denotes a higher degree of fertility, according to Drayson's 35-year study.

Coarse and straight or coarse and wavy hair also indicate the poll area when maximum fertility has been achieved. When thinning or silken texture evolves, the bull falls off his throne as a suitable sire and even descends to infertile status. Nutritional imbalance is the most likely cause of dysfunction, genetics being excused from consideration. Injury is often indicated. Any dysfunction that persists always announces its mischief via hair. Sperm production can literally cease in one year.

More on Bulls

Most of the bulls up for sale fit in the lower categories as calves. This fact is largely unknown because few people are measuring their calves. The usual procedure is to eyeball the animal, to mentally project the calf into adulthood, prepotency being assumed and genetics reduced to buzzword status. Secrets of the bull await discovery with failure and a pasture full of cows that won't settle.

If a bull is not settling cows at 12 months of age, he probably isn't needed except, perhaps, as a terminal cross bull. Bulls or heifers out of such a bull translate into problems. Genetics that are managed and controlled are progressive. Unmanaged genetics are regressive. Both of the above are a given.

The optimal bull at a year of age will have a testicle between six and seven inches in length. In a good year I can identify 20 out of 500 examined. The average scrotal circumference should be between 38 and 40 centimeters. This defines a small window of acceptance, the abnormal sperm cell that inhabits that animal is telegraphed by the dimensions of the scrotum. A scrotum of less than 38 centimeters, or less than six inches in length, reveals an abnormally high cell count. The difference between a 10 percent and a 20 percent abnormal cell count staggers the imagination, and the pocketbook. The single revelation is open cows.

The top sire as defined by Drayson will get 90 percent of the cows pregnant in the first 21 days of the breeding season. Such a bull might copulate as many as ten times in a day because of the testosterone he's producing. The next runner-up probably is incapable of intromission more than eight times a day. Moreover, the abnormal count is higher as the ladder of excellence is descended. The pattern is clear. When the bulls measure up in terms of the

Drayson data, all the cows become pregnant. It is never possible to argue with signs, symptoms and lab results. When semen analyses are low and abnormals are high, all conspire to confirm what scrotal dimensions said in the first place.

Without an outstanding bull, there is no chance at all of producing an outstanding bull calf. Nature provides only a small window of opportunity to create good genetics.

The live count of 75 to 90 percent puts an 80 to 90 percent conception on the other side of that ever-present equals sign. When a bull reaches age eight, the conception rate is bound to go down.

The records illustrated in the boxed material on the following pages show how the sperm count has increased. The animal is capable of producing 300 or 400 straws of semen a week. The live count would be extremely high before extenders are added and freezing is invoked.

If you have a bull or buy semen on the top bull, you can expect somewhere around 60 to 70 percent conception via AI. Frozen semen from the next category down bull will result in five to 10 percent less pregnancies. Semen from the *unacceptable* bull will have the cattlemen pulling their hair out. The AI program won't be working.

AI Programs

There are several AI programs that do not work. The quality of the semen comes first. Invariably the industry has never paid much attention to the research supporting what these paragraphs reveal. The semen is 70 to 75 percent live and it will make the semen with about 25 to 35 percent live cells. They sell it that way.

Earlier, I related how you can observe the hair on a bull before you pull semen to see whether he will pass the semen test. Hair is controlled by hormones. This really is suggested here because the bull in the pasture is the cash crop for the next year. The absence of a club-like tail is a warning flag. The defective tail

Dimensional Scrotal Measurements

Classification	Length (inches)	Circumference (cm)	Sperm Count per cc (range) $x10^6$	Approx. % Live	Approx. % Conception
		Age: 7 1/2-9 months			
Optimal	5, 5 1/2	28, 29	N/A	N/A	N/A
Tolerable	4 1/2	26, 26 1/2, 27, 27 1/2	"	"	"
Objectionable	4	24 1/2, 25, 25 1/2	"	"	"
Undesirable	3 1/2	23, 24	"	"	"
Unacceptable	3	20-22	"	"	"
		Age: 12-16 months			
Optimal	6, 6 1/2, 7	38, 39, 40	980-1379	75-90	80-90
Tolerable	5 1/2	36, 37	672-1076	65-70	70-75
Objectionable	5	35	527-707	55-60	60-65
Undesirable	4 1/2	34	362-538	50-55	45-55
Unacceptabe	4	30-33	40-372	10-45	5-40
		Age: 16-24 months			
Optimal	7, 7 1/2, 8, 8 1/2	40, 41, 42, 43, 44	1093-1790	75-90	80-90
Tolerable	6 1/2	37, 38, 39	1043-1592	65-70	70-75
Objectionable	6	36	796-1541	55-60	60-65
Undesirable	5 1/2	35	381-1093	50-55	45-55
Unacceptable	4 1/2, 5	30-34	309-783	10-45	5-40
		Age: 24-36 months			
Optimal	7, 7 1/2, 8, 8 1/2	43, 44, 45, 45 1/2	1379-1853	75-90	80-90
Tolerable	6 1/2	39, 40, 41, 42	920-1469	65-70	70-75
Objectionable	6	37, 38	732-1181	55-60	60-65
Undesirable	5 1/2	35, 36	517-1011	50-55	45-55
Unacceptable	4 1/2, 5	30-34	68-548	10-45	5-40
		Age: 36-48 months			
Optimal	7, 7 1/2, 8, 8 1/2	43, 44, 45, 46	1218-1990	75-90	80-90
Tolerable	6 1/2	40, 41, 42	965-1790	65-70	70-75

Classification	Length (inches)	Circumference (cm)	Sperm Count per cc (range) x10⁶	Approx. % Live	Approx. % Conception
		Age: 36-48 months (cont.)			
Objectionable	6	38, 39	809-1423	55-60	60-65
Undesirable	5 1/2	36, 37	559-1043	50-55	45-55
Unacceptable	4 1/2, 5	30-35	104-592	10-45	5-40
		Age: 4-5 years			
Optimal	7 1/2, 7, 8 1/2	44, 45, 46, 46 1/2	1060-2066	75-85	80-90
Tolerable	7	42, 43	863-1567	65-70	70-75
Objectionable	6 1/2	38, 39, 40, 41	486-1296	55-60	60-65
Undesirable	6	36, 37	409-732	50-55	45-55
Unacceptable	5 1/2	30-35	54-570	10-45	5-40
		Age: 5-6 years			
Optimal	8, 8 1/2, 9	46, 47, 48	1163-1730	75-85	80-90
Tolerable	7, 7 1/2	42, 43, 44, 45	783-1618	65-70	70-75
Objectionable	6 1/2	39, 40, 41	507-1060	55-60	60-65
Undesirable	6	36, 37, 38	309-863	50-55	45-55
Unacceptable	5 1/2	30-35	47-836	10-45	5-40
		Age: 6-7 years			
Optimal	8, 8 1/2, 9, 9 1/2	46, 47, 48, 49, 50	1218-1790	75-80	80-90
Tolerable	7 1/2	42, 43, 44, 45	877-1423	65-70	70-75
Objectionable	7	39, 40, 41	592-1337	55-60	60-65
Undesirable	6 1/2	36, 37, 38	362-836	50-55	45-55
Unacceptable	6	30-35	40-695	10-45	5-40
		Age: 7-14 years			
Optimal	8, 8 1/2, 9, 9 1/2	46, 47, 48, 49, 50	920-1469	70-75	70-80
Tolerable	7 1/2	42, 43, 44, 45	707-1423	60-65	60-65
Objectionable	7	40, 41	548-892	55-60	55-60
Undesirable	6 1/2	38, 39	194-683	35-45	30-40
Unacceptable	6	33-37	40-507	10-30	5-25

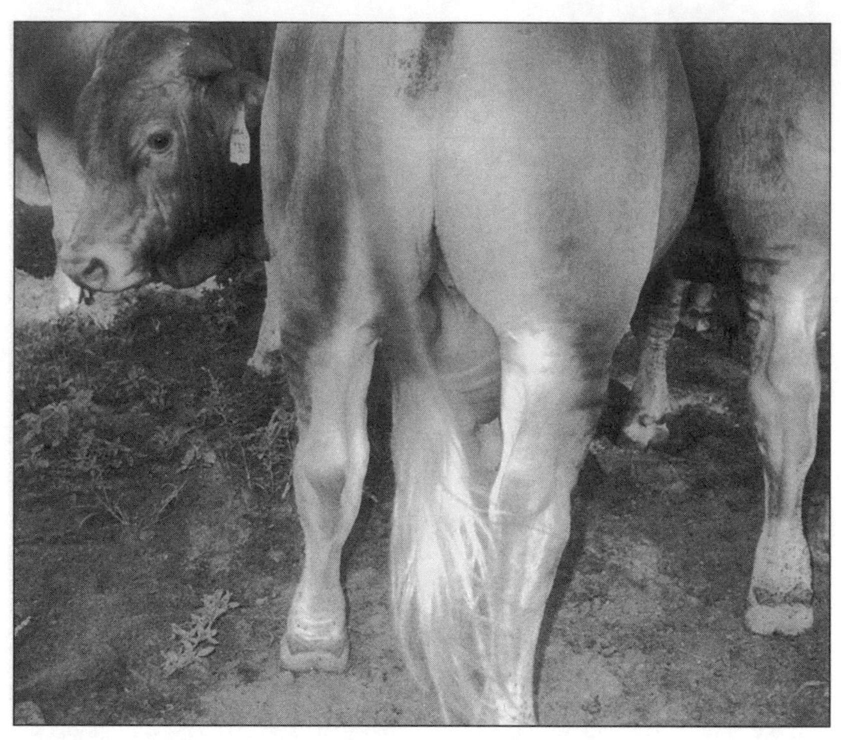

Signs and symptoms can be read on the tail structure.

structure is indicated by small bones and silky hair. Signs such as these can be read from a pickup truck at the edge of a paddock.

Cell Biologists

Cell biologists now tell us that there are areas of the body where the cell wall can become thin. The udder and the testicle area are two places where this occurs. When deficiencies occur in the animal, the cell wall becomes thin, allowing white blood cells to flood through and invade the seminal fluid or into the milk duct area and coagulate, as in mastitis. This sets up an infection.

Five-way and seven-way vaccines kick open so many doors to mischief, they never permit simple answers. Injections, minerals, vaccinations, all of these are Band-Aids. Genetics are the real problem and the real solution.

At the point of conception everything is determined, reproduction, maintenance, udder, bone frame, feet and legs, ability to pass traits, hair color, temperament. It is this inventory that cancels out single trait selection. At the point of conception, the pituitary is determined. The pituitary controls the adrenal gland. A swirl of hair designates a robust adrenal, as previously illustrated.

The liver stores nutrition. Trace nutrients such as copper, zinc, manganese, molybdenum, selenium, etc., have to be processed in the liver and stored for use on demand. The wrong form of minerals in the animal can overload its system, kill an animal and have a tantalizing effect, copper deficiency. The liver governs, stores, rations. The external sign of a functional liver is a bright, shining coat.

The Thyroid

The thyroid is situated under the back of the jaw and neck. Bonsma tells us that a bull or cow with long hair on the neck has a thyroid problem. That animal is not secreting hormones and enzymes that govern shedding. Moreover, that same sign suggests a fertility problem. The pancreas gland, in bull or cow, is located on the side, straight up from the navel. Experience often comes to the rescue in ending fertility problems.

The testicles and the crest on his neck are signs and symptoms, to stretch the last word. If they appear and mature at an early age, then there is a pituitary gland that is developing early.

Morphology is determined by complete interaction between heredity and environment. External affecter organs, skeletal size, muscling, fat deposits, hair coat, color and breed, all are choreographed by the morphology of the animal. The animal that does not shed identifies its own problem, a gland system that is not functioning. Genetics or management can be the culprit, the animal the victim. Management almost always means meeting the requirements of the animal in terms of food, water and comfort.

If your objective is to breed two breeds, say a Brown Swiss and a Jersey, failure to use a highly masculine fraternal bull or a highly feminine maternal cow will result in disappointment. The problem in the dairy industry is finding either. According to my attempt at inventory, there are not many out there. If you could find the kind of cow and bull I look for all year long, you would have an animal that could harvest grass most efficiently.

I'll use a Black Baldy to illustrate the same point in beef production. It is a cross that works. It really should be an end product but most farmers keep them and involve a third cross. Three crosses in one animal usually results in a 60 percent loss of efficiency. I can illustrate the point with at least one case report. A client with 100 Black Baldy cows bred them to a Charolais bull. The result was about 30 to 35 percent outstanding calves. Such progeny commands attention. Lost, much like a sapling in a mature forest of trees, are the 60 to 70 calves that are mediocre or less.

When you out-crossbreed or crossbreed and the progeny does not out-perform either of the parents, the procedure cancels itself out. The breeding program is not working.

Cowman Conundrum

Calving problems are born when the breeding match is made. In a lot of Holsteins, for instance, hook bones and pin bones are level with each other. This means a square-rumped animal. It is not uncommon to see them higher. But there should be a two to three-percent decline from the hook bone down to the pin bone. All the other measurements mentioned earlier seem to issue their orders during gestation as noted. None can be changed once issued. Genetics has already decided the issue and it can be indicated only before conception is achieved, selection the tool.

Breed magazines and TV footage try to identify the bull as the determining factor in achieving herd excellence.

Large-framed females tend to be out of balance and manifest low fertility and paternal characteristics. They go a long time between calves and are hard to settle. Maternal traits cannot be wished, they have to be found. The right signs are in the rump, in the flank. This can be observed and then confirmed by measurement. The absence of the deep, wide rump starts out as a classic mistake. If the proper dimension is not achieved, that calf will not be a maternal cow. Most probably, that animal will be late maturing, will not be able to mate with a masculine bull and reproduce a calf with replacement qualities.

Heifers with a plus six-inch flank measurement larger than girth have meat all the way down to the elbow and the back leg. They turn out to be outstanding cows. The difference in heart girth and flank measurements is correlated to increase in maternal ability and red meat production. Bone with no meat adds to maintenance without return.

A Practical Matter

As pointed out earlier the bull should be 44 percent as wide as he is tall. If he's not, the cowman is giving up a volume of meat. I like my bulls to be 47 or 48 percent. A 50 percent ratio is hard to achieve, for which reason I point to that British White bull mentioned earlier and later in this text. Male hormones correlate with the ability to gain efficiently. Animals with width to height ratio at less than 44 percent tend to be out of balance. An absence of meat in this area nears a similar absence elsewhere, as indicated by a tight heart girth or a narrow front end. A failure to achieve 44 percent indicates a late-maturing animal, low-quality semen and a lowered carcass cut out in steer progeny.

Breeders with years and experience under the belt have short-circuited the need for one measurement for generations. They noted correctly that when an animal walks, its back foot sets down in the track of the front foot. Sometimes the animal oversteps it. That walk indicates the cow can have a calf of the breed of normal size without any problems. There is no structural defect. If there is a structural defect of any kind, the animal will not track right, and there will be a short step.

If a bull is post-legged, meaning the leg fits into the pelvis half an inch or an inch too far back, then that leg always will be straight. That hock won't have a set-in; it will be straight. The faster the bull walks, the shorter the step and the greater the inability to track.

If you measure from the ground to the animal's back — the stifle muscle is in that area — that's the thurl measurement that needs to be at least 13 percent of the total height, a 14 or 15 percent readout is better. This measurement, much like simple observation, governs the ability to travel. A bull that can't travel is always marginal.

A Bull Art Form

Michelangelo's art form was the human body; his well-muscled statue of David remains a classic form centuries later. One can be allowed to wonder what process of observation the artist might exercise were he to find a superb bull within a herd. My guess is that he would anoint the head with the coarse hair that denotes virility or fertility regardless of the breed. The peach fuzz texture of the pansy poll is not a signal of potency, and breed is merely individual, not a commercial factor. The hair shaft population seems to find its genesis in hormone and testosterone strength, not in the comb and clipper of the dilettante. Hair follicles and population are not physiological objectives, thus the moderate amount on the top class bull.

Coarse and wavy hair on the face and head denotes a higher degree of fertility than the owner of coarse straight hair, according to Drayson, whose observation I definitely confirm. There is still a higher degree of fertility as engraved in the hirsute stand about the face and head, the coarse and tight, curly hair. The curl and tightness correlate with the peak of sperm production. Records put this peak at ages three to seven. The old thumb and forefinger test should suffice. Simply straighten out a curl and watch it return to its clock spring posture.

Images of bulls

Judges of bulls do not note or remember this quite simple sign because of blockheaded handlers clip and comb the Samsonesque badge into oblivion, or they ignore its implications. Ignorance decrees that fine hair complements the requirements of a shiny photograph because a fine texture blends more easily with the air brush for presentation in the annual bull book.

A fine texture is not fatal to the objectives of the seed stock producer. Quite the contrary, a fine texture means strong sub-fertile tendency. A robust hair population tells its part of the story. It proclaims that there will be no rapid change in either hair pattern or fertility status very rapidly. Some three to six months will elapse under the best care and nutrition to effect the desired transformation.

The claim between acceptable and non-acceptable bulls is further described by hair population and coarseness in the poll area. The coarse and wavy or coarse and straight rule governs. When that clock spring hair stands up and loses its tension, a physiological change is taking place, one that will demote the bull from fertility to sub-fertility. Astute observers will note a change within three days after trauma or disturbance in bodily function, often a consequence of nutritional imbalance, possibly a change in ration quality. If the condition worsens, the poll hair will send up its signal, fertility going into further decline. When young, girlish boys appear, the bull is in serious decline. The eventual goal of degeneration is sterility. Blunders have a way of registering themselves — usually within a year — moving from excellence to sterility.

The full roster of ailments with temperature producing potential that are discussed in the chapter on nutrition invites consideration. Hair follicles on the rampage suggest the need for professional help.

Neck hair also denotes fertility or lack thereof. Again, it denotes hormone power and the ability to settle cows. Breed does not seem to matter, not when neck hair is clock spring curly and resilient. A natural sheen of hair should anoint the entire body. When neck hair becomes straight and kinks less, decline has taken place or is imminent. The appearance of fine neck hair confirms what hairs around the face and neck have already declared — the steady dive into sub-fertility.

Certain beef breeds have horns, of course. Changes in horn presentation take place during a bull's lifetime. They denote age, maturity and fertility. Early on the road to maturity, horns are creamy white. At puberty horns take on a rosy-red color, meandering out from the base of the horns. At age 14 to 15 months, horns should have achieved that rosy-red color all the way, or at least two-thirds out. Once colored out, the horns will retain that hue until three or for years of age. It is then that the horns change to gray or olive.

All this to warn that mottled horns are a red flag. Sub-fertility can be confirmed and it always can be evaluated under microscope. Sperm cell mobility is frequently observed and a low sperm count can be expected. Horns tell you, if you care to observe, that testes are not able to produce live sperm cells.

Often white rings will be observed on horns. These denote sub-fertility or sterility because of some impediment, a nutritional imbalance, trauma, or an inherited genetic defect. The white ring appears quite suddenly in the wake of an impediment's arrival. Here removal of the impediment, once discovered, is followed by less evidence of the ring. The ring will remain, offset somewhat by the normal color. After the problem has been converted, a chalky-white set of horns means a sterile bull, one that can no longer settle cows or will do so only occasionally. Such horns will be accompanied by the hair signs noted earlier. Semen analysis will confirm, never deny, the above. Once a bull achieves sterility, some recovery is possible, but usually it is a lost cause.

Horn coloration is largely derived from the color of the breed. Blacks will have black horns. In the case of blacks, mottled, or gray rings amount to color. The hair signs remain the same. The bull as an art object has lines and ripples that summon the hammer and chisel, were one to capture such an animal accurately in stone. Prominent on the fertile bull is the neck vein. It either governs or identifies the androgen hormone production in the testicles. This vein grows gradually. It "hits the ground running," to use a figure of speech, when the animal is young. It grows as if to identify puberty and flares out as sexual maturity is achieved. Compared to a neck vein on a steer, this river of blood is mammoth.

Should sperm production be reduced or lost by the bull, the neck vein will reduce itself in size. If the neck vein muscle flattens, the masculine look will almost disappear. The area behind the head will shrink three to four inches. This loss delivers a shock to those who compare the bull with a second bull not affected by an interference with his fertility. Decline is always slow, requiring 12 to 15 months. It takes an alert cowman to take notice. This is why measurement, observation and intense scrutiny remain the currency that buy the top status in fertility.

The Tail

I have mentioned the tail of the bull before. For the sake of continuity, a few notes are now in order.

The rule of hair prevails. Its coarseness is a given. As the tail descends, its thickness diminishes but remains solid. The old wives' tale cites a hollow tail and calls for an injection of everything from kerosene to penicillin. It need not detain us. There is no hollow tail. There is a slim tail with a refined appearance, more a female tail than the robust tail of a bull. This denotes sub-fertility.

The scrotum is one of nature's wonders. It is air cooled to less than the body temperature where spermatozoa are stored. The bovine has a body temperature of 101.5 F. The scrotal temperature is 98.6 F, a 2.9 degree variance. Nature has set up this requirement to permit sperm production, development, maturation and storage in the epididymis. The scrotum is elastic. This pendulous bay contains the dartos muscle under the skin of the lower half of the scrotum. It has a prime role in maintaining that

98.6 degree temperature. The dartos muscle contracts in cold weather to tap added body warmth for sperm preservation. In hot weather it reduces to permit these air-cooled testicles to distance themselves from excess body heat. Environmental temperature often assails the scrotum's position. Frost damage to the scrotum is common. A shortage in fat protection is a hazard when breeders seek to lower the fat amount of animals via single trait selection. Such animals, moved from a warm climate to the northern tier of states often will not make it through the winter. The prize bulls are often the victims.

When severe frost damage occurs, the epididymis becomes damaged and this damage impairs a bull's ability to produce semen. Epididymis damage is often a prime cause of sterility. Basically, the epididymis is a nutrition and storage facility. Since delicate storage of sperm cells is required, damage of the warehouse is tantamount to damage of the product.

Maintenance of a clean scrotum is next to godliness. Caked mud, untreated scratches, frost damage, all can assist in the incitation of infection. Bacterial or fungal invasion of the area near the epididymis can have grave consequences, the worst being sterility on low sperm production preservation. Infection means temperature higher than the normal 98.6 degrees.

Scrotal hair is not without interest. It should not increase in winter. Unfortunately, it often increases in the winter, especially in mature bulls. It should be terminally short or totally peach fuzz, as in a buckskin texture. Hair two to three inches in length follows the pattern observed elsewhere — it is an indicator of low fertility. Long silky hair is a consequence of low fertility. Its very presence means a scrotal temperature of over 98.6 F, reduced semen production, a shortfall in maturation. If long hair is shed, fertility could exhibit recovery.

Scrotal hair can be read like a road map even as puberty of a bull calf is achieved. If the long scrotal hair of youth does not vanish as puberty is reached, that bull calf will not achieve high fertility. If this failure to shed and replace is not evident at six or seven months of age, heredity can be blamed in the absence of sickness or injury. If 14 or 16 months of age does not remedy the situation, the condition must be considered permanent.

The actual measurement of testicles now comes into focus more clearly than in the discussion of dimensions in Chapter 3. Both testicles should be carbon copies of each other. This means dimensions, degree of firmness and shape. When these three factors clearly synchronize, weight and volume can be assumed. The objective is uninhibited sperm production.

Conception is the clear objective; it is and remains the bull's job to simply settle cows, and scrotal excellence is his badge of office. But there is more to the scrotum. It serves up libido, or sex drive.

A Reason for Being

Each testicle has its reason for being. It is made up of an extremely intricate system of connective spermatogenesis tissue. Clinically, I can simply say that sperm-forming tissue are located in the somniferous tubes of the testicles. These are oil-like structures that have been computed to being 10,000 or more feet in length, surely one of creation's wonders. How these sperm cells are formed can be described in the sterile terms of science, yet we have to hang our hats on the simple statement, "we don't know." Laboratory people muck around with genetic engineering and support changes based on DNA manipulation, all being produced without very much understanding of what we already know. Suffice it to say we know how nutrition figures, how genetics rule based on the clear understanding that no one can manage genetics to withstand starvation; those tissues and organs do their job when we have the wit to manage a level playing field for the animal.

"Primary sex cells of the male grow and divide to form primary spermatozoa," wrote James Drayson. He went on to detail how cell division takes place. As the division goes on, spermatozoa result from the intervening cellular delivery. The route through tubes takes some 33 to 38 days, at which point sperma-

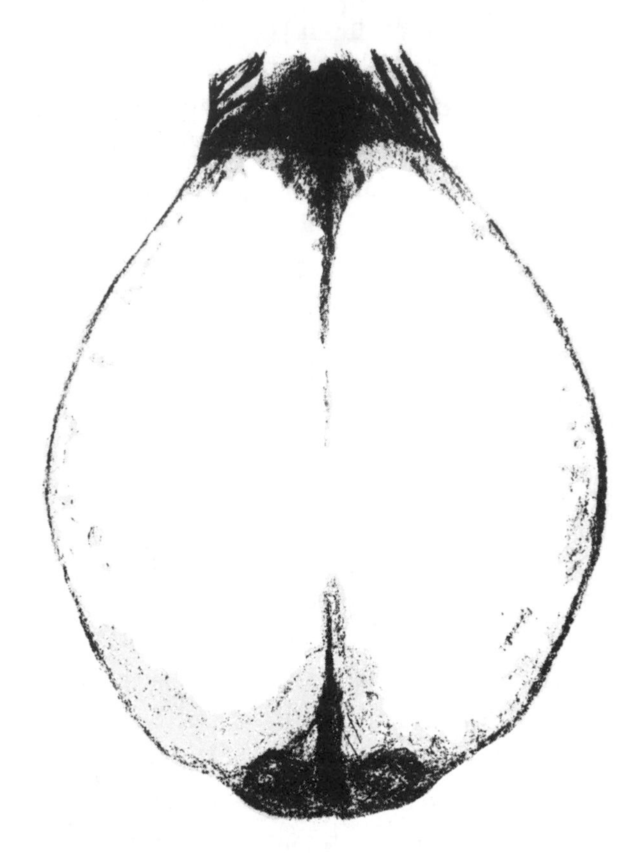

The normal shape and approximate half size of a set of testicles of an optimal bull. (Illustrations on pages 82-85 are based on diagrams from James E. Drayson's Herd Bull Fertility.)

tozoa pass into collecting ducts to the epididymis where the material finally achieves mobility and maturity, awaiting ejaculation. Thus fertilization of the ovum.

The system seems flawless and yet many things can interdict fertility, performance, and results. Stress, general health, mineral imbalance — meaning a shortage or marked imbalance — all have been indicated. Seldom mentioned is the lack of knowledge, especially among dilettantes, amateurs and pseudo-intellectuals.

There is art and science in the business of examining a bull, and there are findings only diligent work can reveal. The epididymis is a point in question. It will vary from the size of a pea

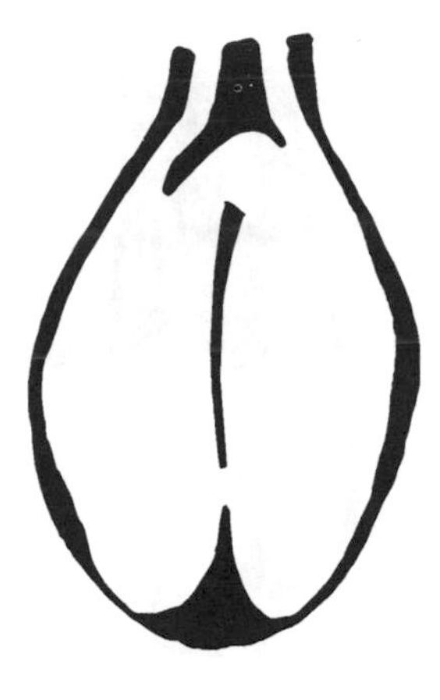

Normal

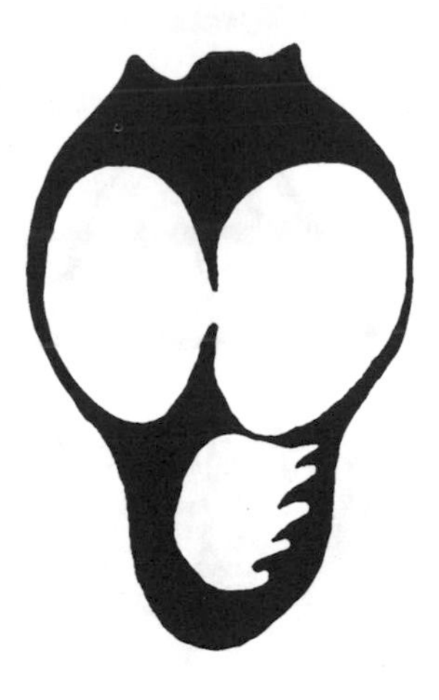

Pear degenerating
bottom to top

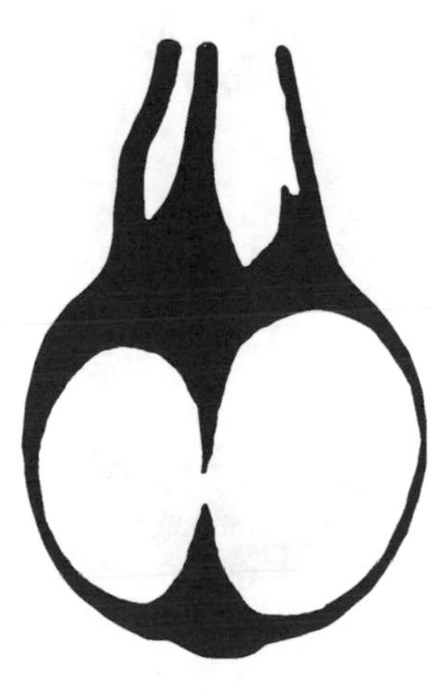

Pear degenerating
top to bottom

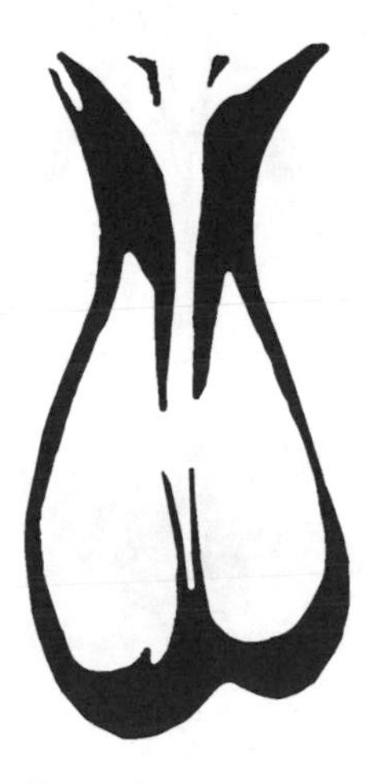

Prominent cord

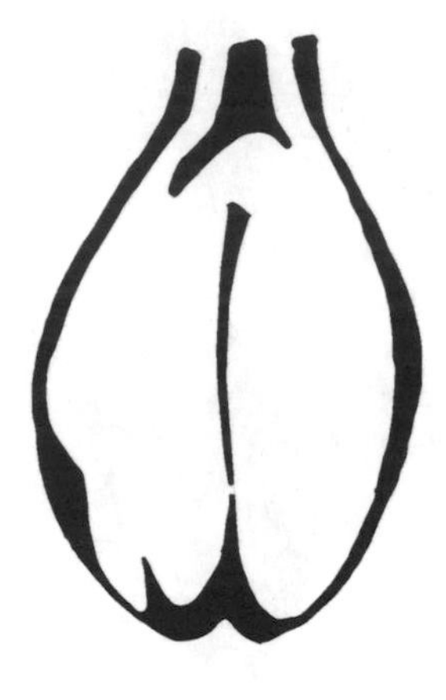

Inverted

Cylinder shaped

Epid Large

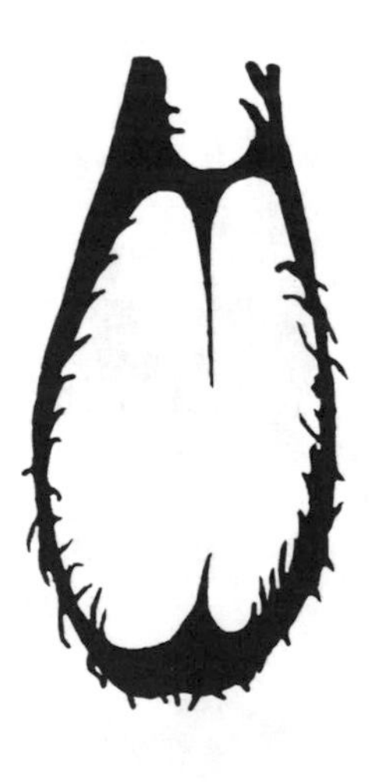

Epid Small

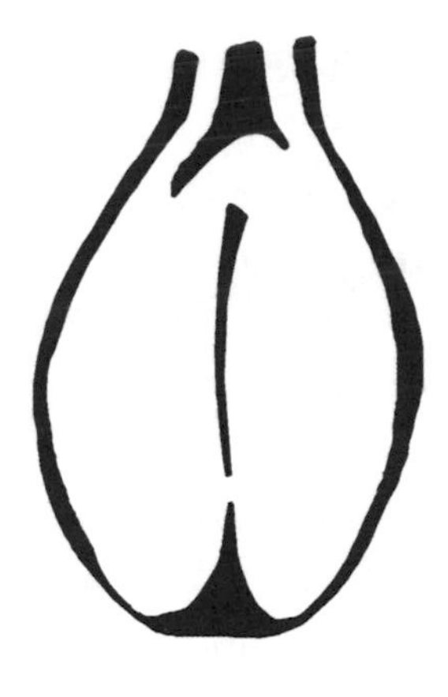

Flattened

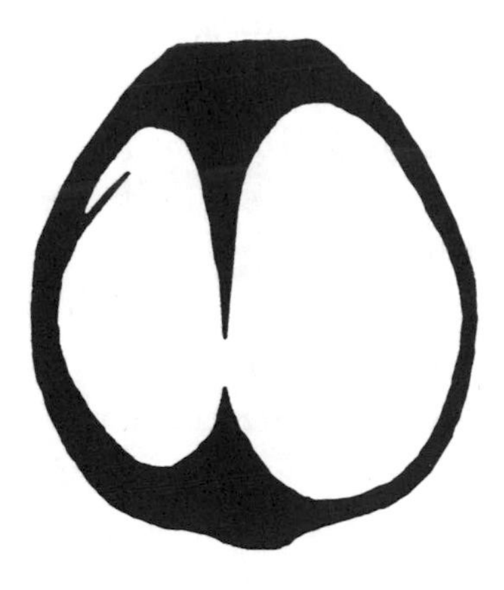

Grapefruit

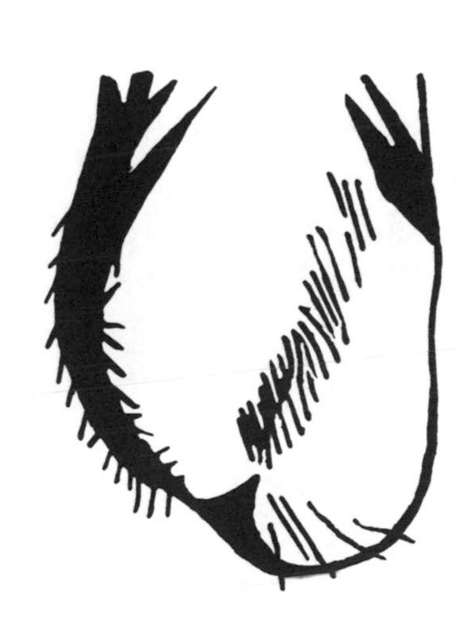

Left 1/2" shorter

Mushy

to the size of a golf ball. The first is too small, the golf ball size too large. It should be about the size of a pecan.

Withal, testicular degeneration spells the end of a bull. Loss of shape and size is the harbinger of descent into sterility. Shrinkage of size is the key index, but the reverse also can be true and the testicles can become enlarged beyond the optimum, which also suggests degeneration. A 6 to 10 cm increase in size poses and answers the question; Drayson's chart on optimal dimensions is not without interest.

Testicular degeneration is often a consequence of excessive use, injury, shipping stress, and the defects discussed earlier. Palpitation reveals a lack of firmness as degeneration proceeds. A mushy testicle texture seldom can be repaired. A prime cause of degeneration is the use of drugs, vaccines, and estrogen loaded feeds. It almost always takes four to six months to overcome the use of drugs, whatever the reason for use. Reproductive capability is reduced 50 percent by drugs.

A Codicil of Sorts

I have to close this chapter with a codicil of sorts. It came to me while on a cow-buying trip to New Zealand on behalf of a client, the New England Livestock Alliance. They wanted heifers of a quality only New Zealand could supply. Matched with semen from quality Devon bulls via AI, the expected result is a recovery step for a breed that has been allowed to deteriorate over the last few decades.

My host was a line breeder, Ken McDowall of Rotokawa Estates Ltd., New Zealand, who kept his bulls and cows on grass, using few if any of the chemicals that scar bovine production almost everywhere around the world. The New Zealander has compared fertility of bulls on grass to counterparts on feed pellets, flakes and manufactured feeds. The grass fed bulls exhibit a live sperm cell count of 90 percent. The preserved fed counterparts often clocked in at 75 percent, usually at 60 to 70 percent live. These last stated data correspond to the live cell count generally available in the US.

The New Zealand findings have been confirmed by Al Williams of the University of Mississippi. Williams was in charge of a bull checking station for nine years. His observation was that bulls performed in the ranges stated above, according to whether they were on pasture or not.

Chapter 5

Classic Cattle

A Curse Is Never Causeless

Having mastered the art of measuring as taught by Bonsma and Winchester, and the art of observation suggested by Drayson, I was prepared to enlarge my knowledge when I met Lester Lodoen, Bruce Campbell, Doug Paton, and Ewan Clark in Saskatchewan. Lester Lodoen is a farmer and resident of Canada. Bruce Campbell, Doug Paton, and Ewan Clark are from Australia. Their observations and grading system invite attention because it is the key to understanding the taste and texture of meat protein, a benefit sadly compromised and often missing in commercial red meat.

In terms of genetics, it is a given that the bull and dam share equally in delivering traits to the offspring. That much stated, details generally are regulated to archival status and forgotten. But the four gentlemen I encountered had been studying bulls and body conformation of cattle for about 20 years. Their findings have been codified into two manuals styled *Classic Livestock Management Evaluation,* which are generally available from

Lester Lodoen, PO Box 127, Fox Valley, Saskatchewan, Canada S0N 0V0.

The four named above have added to the inventory of information contained in the precis of these chapters not only in how a bull affects progeny, but in how observation can be dovetailed to measurement for herd progeny, both male and female.

In the *Classic Livestock* encyclopedia, the makeup and phenotype of the scrotum, complete with udder conformation, principal conformation and with predicted fertility of the animal a bull will produce are described. A scheme of rating based on a one to five score has revealed that 50 percent of udder formation is related directly to the scrotal makeup of the bull.

Bulls have false nipples, much like a male of the *Homo sapiens* species. These nipples are located on the neck of the scrotum. When a bull sires a daughter the offspring will have a tilted udder with the front quarters being very light. These bull nipples never grow and for practical purposes appear to have no function. They are supposed to be in front of the scrotum, or on the flat part of the belly, forward and an inch in front of the scrotal sac. If these nipples are displaced, detection of this fact telegraphs disturbances of every kind in the daughter's teats.

The culprit is genetics. The genetics of the scrotum affect the genetic makeup of the udder. Such bulls are not feminine, nor are they heavily masculine. They have simply lost their claim on perfection. They are undersized in both circumference and length, for which reason any son will not be very masculine. Down track sons produced will not be very masculine. It would be exhilarating to suggest that the above observation flowered full grown, as from the head of Zeus, but that would be less than accurate. A legendary old man in Australia passed on his oral testament some 20 years ago. An old timer in Canada confirmed the same thing to Lester Lodoen. Unfortunately, pseudo-science perplexed a great deal of native knowledge over the past 50 years, interdicting its continuity and making transmission to future generations difficult.

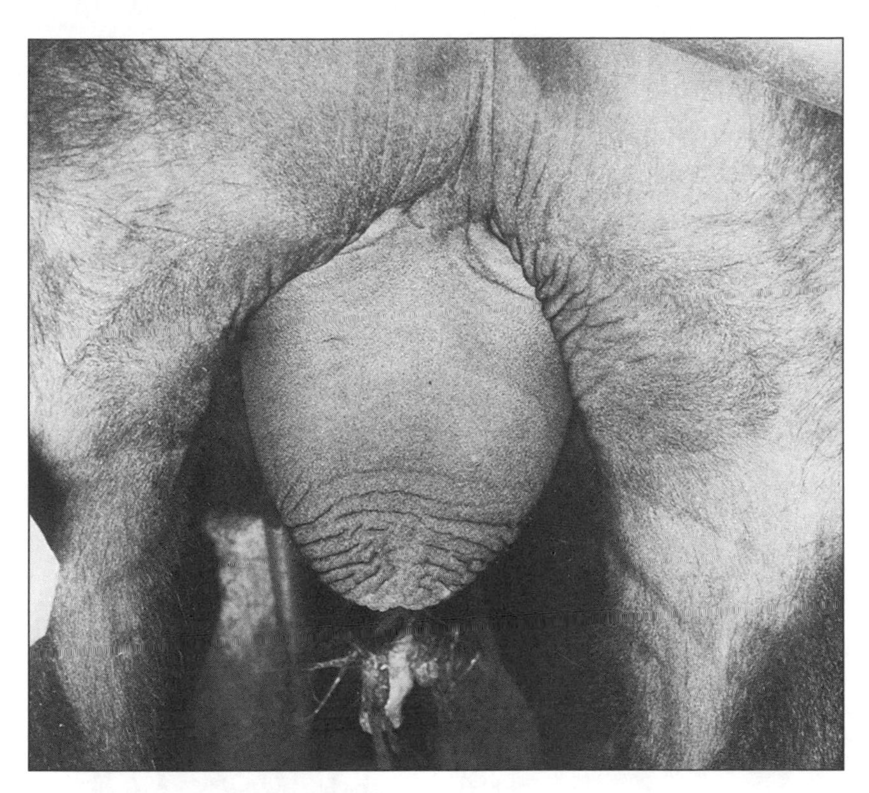

Scrotal makeup of classic bull.

That this body of knowledge coincides with what I have been unloading on cowmen over the past several years goes without saying. Reproduced is a photograph of a bull that will produce a perfect udder in the offspring. The scrotal makeup is represented above.

Breeding According to Traits

The Canadian-Australian-New Zealand triumvirate has identified no less than ten traits that conform and compliment the measurements discussed in the first few chapters of this book. They are angularity, basic form, foreleg shape, body components, femininity, calving capability, udder, teats, reproductive capacity, hormonal activity and temperament. Most of these have been identified by tape and ruler, leaving room for the practiced cowman to eyeball every measurement revealed or implied. Some traits, such as flatness of bones, are examined by hand. Examination of the lubricated patch along the spine require both touch and vision.

The full scope of this discipline does not fall within the purview of my outline. I merely mention it here so that readers pursuing knowledge no one book can cover will find elaboration in the study course the *Classic Livestock* manual has to offer.

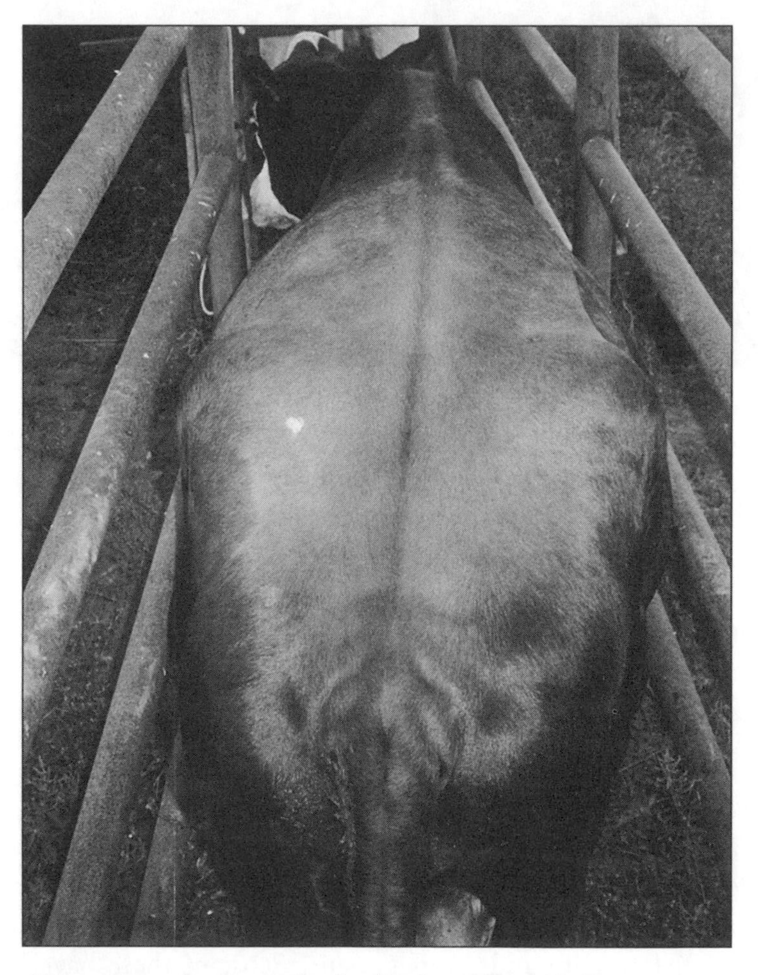

Illustrating the sign, the back of an animal that promises palatability.

Two of the classic assessments are for palatability and tenderness. The flatness of an animal's bones is an indication of tenderness. An accurate assessment can be found in the flatness of the jawbone. Palatability, in turn, is related to hormonal activity. The skin of the animal is the index and the shining strip running down the animal's spine (see photo on page 92). Feedlots ignore tenderness and palatability. Five star restaurants do not if they intend to remain five star. The influence of femininity in allotting the appropriate taste and palatability level should be noted.

The tone angularity — much as with an artist's conception of the horse — is central. The single concepts in making the appropriate determination are feminine features, male testosterone, fat distribution, maternal rump, and thick flank and stifle muscles. There should be a V-slope from the neck to the hind quarters, this from a top view and side view. The spring of the ribs — all this to say angularity indicates fertility and longevity. These few comments illustrate the general approach of the Canadian-Australian team I encountered in Saskatchewan. The running comparison to my measurement teachings should be at once apparent. A new dimension, one I have not stated or elaborated, it has to do with testicles and udders, positioning of both being a part in question.

Scrotum and Udder

It will be noted that the line drawings of the bovine scrotum in Chapter 4 do not include placement or identify the position of small seemingly useless nipples on or near the attachment of the scrotum to the body. This was not an omission. It allowed me to defer remarks pertinent to the outline of this chapter.

I will start with the observation that in the female the vulva should be straight up and down. This seemingly out of place observation is included here because it asks a question about high pin bones and a consequent poor drainage of the cervix. High hips and thurl create a pressure that reduces the overall pelvic area. A muscular rump often results in hip lock during calving.

A horrible example of a pendulous udder.

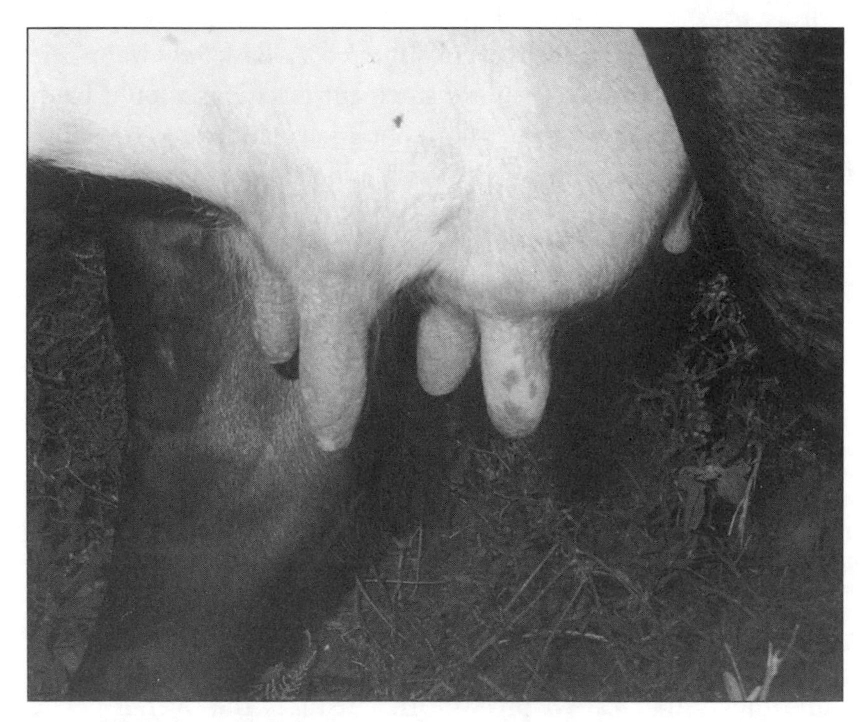

A well-balanced udder.

This now stated, it is now appropriate to further analyze both udder slope and testicle size. Placement of the udder too far back allows the udder to stretch forward and halfway up the back toward the tail, a requirement for maximum milk capacity. This suggests a continuous curve from the belly to the tail, a pendulous udder strains the connection. The oversized udder, with teats swinging so low they invite injury by stepping, is no asset. Moreover, the udder should be high enough to permit unrestrained sucking by a calf. I have included here two photographs of teat locations and examples of teats that invite infection and prevent a calf seeking nourishment.

Balance is the thing cowmen should seek. A hindquarter should be of moderate length, width and depth. The depth should relate to the hock, capacity and clearance a given, the floor about five centimeters above the hock.

The female of the species has as much to say about the progeny as the bull. Of course, a cow has only one son or daughter a year, whereas the bull can have between 50 and 60 sons and daughters. Collectively, the female balances the bull's vigor. There are impediments, for instance, navel weaknesses can lead to loss of a calf. Both poor mammary stimulation and closing of the mammary prior to calving suggest a time bomb. Thus the role of the calf in hormonal activity for cycling.

The counterpart to the cow's udder is the positioning of the scrotum, its shape, attachment and size. The scrotum should be positioned well back between the hind legs. The photograph on page 91 is an example of a position best calculated to support conception in the cow and hormonal balance in the offspring.

The perfect football shape with the epididymis at the very bottom of the testicle will be noted. A segment line from the bottom to the top between the two testicles illustrates a slight cleavage. On a distorted scrotum if the testicles are not made right and the epididymis are not on the very bottom of the scrotum, then that cleavage line is only a fourth of the way up, or sometimes even missing. This causes the udder of the cow to turn out, teats like spigots at an angle, and not perpendicular. In such a case, there is no restricting ligament usually put in place by correct genetics. A defective udder lays apart, it falls down. This is not because of heavy milk production. It is because of the missing ligament.

Not all of this knowledge relies on the testimony of old timers. Bonsma very specifically talked about not having the nipples on the scrotal sac, and he showed that when nipples appear on the scrotum, the prepuce will angle straight down toward the ground and exhibit excessive navel hide, signaling a bull of low fertility.

It all has to do with the epididymis and the scrotal confirmation. Whenever you have a bull with a small or absent epididymis, the resulting heifer either will not have an ovary or have a low-functioning ovary. Genetics will affect the bull or the cow.

Deviation from the norm in the bull always transports its defect to the udder of the cow. The udder breaks down. This results in a large udder, bruises and mastitis — consequences of a pendulous udder.

The teats should be in the very middle of a quarter. Displaced nipples on a bull will result in nipples over to the side or somewhere out of place other than in the middle of the udder. A tilted udder results in low milk production in the two front quarters.

The shape of the udder and the position of teats has implications not restricted to the animal in question. The ability of a baby calf to find its nourishment is severely compromised when the location of the nipple is out of place. A tilted udder creates a V in front. It ought to blend into the belly quite level. Correctly suspended and blended it denotes notches or holes the calf can seek and find. Instinctively, the calf knows enough to hunt a nipple and udder. But when the essentials are misplaced, a probe can result in the mother kicking her own calf. Abnormalities cause a first calf heifer to have a tremendous amount of edema in their udders. This syndrome causes even more kicking because of the inability to nurse.

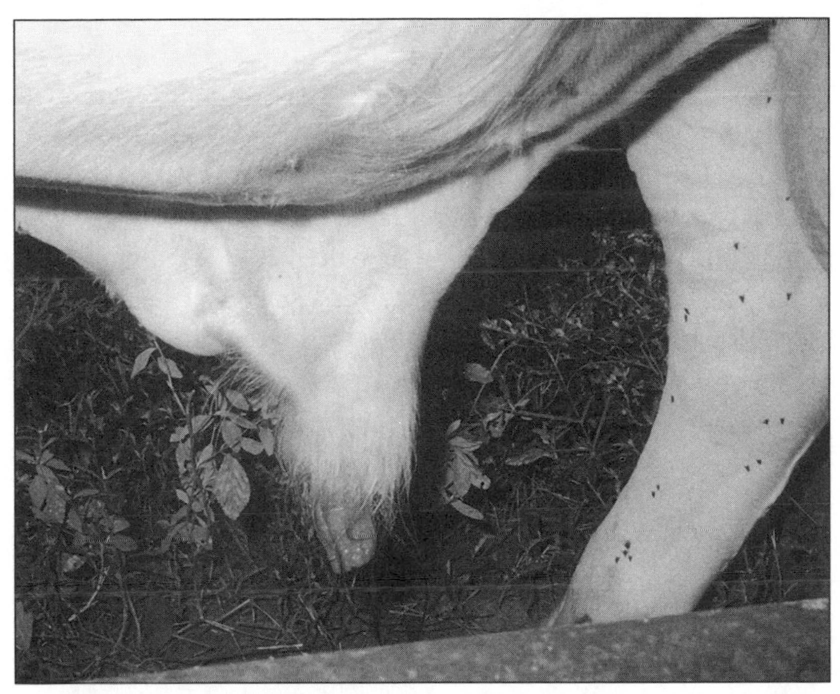

Pendulous scrotum with prepuce hanging towards ground illustrates the inventory of warnings.

A tilted udder.

Implications

Not many in America have equated udder conformation to anything except the cow. I have examined herds from Mexico to Canada. The absence of an epididymis in bulls is always accompanied by a paddock full of overly large udders.

A seed stock breeder I know has been line breeding for 35 years. He has a fence full of cows. Almost every cow he has is ruled by large, pendulous udders. With the restraining ligament gone and teats turned out, it cannot be otherwise. This situation tracks back to the bull.

The bull without an epididymis, on exhibiting bad testicles, will deliver barely 100 straws of semen a week. Any bull that can't deliver 500-600 straws of semen a week is not worth collecting.

Unfortunately, American cattlemen pay little attention. Showbiz looks at the animal from a Hollywood point of view, all show and no substance.

In the U.S. and North America, the qualifications for good straws of semen is 25 to 30 million live cells after the straw has been frozen and unthawed. In New Zealand the semen often has between 700 and 800 million live cells per straw. Genetics and grass generation after generation defined the animals I imported from New Zealand. The bulls to which I refer have all looked like a bison up-front, extremely deep and thick with a deep and wide rump to match.

I have a customer in South Dakota who has been buying EPD bull semen for 25 years. He's created a herd of cattle that is extremely high maintenance. He feeds a lot of his own cattle. He averages four percent less meat per carcass than neighboring counterparts who claim to be at the national average. There is a reason for this. His cattle have light rumps and are light fronted. The bulls that have served up this deficit were recommended by AI experts because they had the best numbers.

The Genesis of Taste

Taste of the meat protein product is governed by hormonal activity. The hormonal activity of the bull governs the hormonal activity of the dam. The glandular function of both is the linchpin.

The thyroid, thymus, pancreas, and adrenal all have to be functioning if taste is to attain the arrival of meat protein on the dinner plate. Basically the hormonal activity also has to do with the quality of testicles and ovaries. If the thyroid, thymus, pancreas and adrenal are not working properly, then the testes and ovaries do not work right. When glands do not function correctly, the animal is subclinically ill. Illness is not always expressed by the presence of fever.

In a manner of speaking, signs are often more meaningful than symptoms. The oily streak at the very top of the back — the area where vets pour phosphates (Phosmet) to battle grubs — is a sign of health. The adrenal and pancreas are high on the back, and it can be assumed pesticides penetrate systemically to deliver more mischief than they solve. An animal that has not shed as explained in Chapter 2, and if the gland system is not functioning properly, the resultant product will have a flat taste. A rich taste and flavor will be missing.

Both the bull and the cow conspire to bestow or withhold taste. That is why selection must consider proper glandular function if taste is expected to be in the offspring. Failure to select a bull with an active glandular system and a cow with an equally active glandular system is to sandbag the progeny in the crap game of life. Taste in the end product won't happen.

Errant Agribusiness

Feedlots use silage, corn, and feeds that add misery to pain. Ensilage feed gives meat and milk a flat taste. A castrated bull fed out as a steer cannot have a maximum taste level. Most steers have long hair atop the neck, and this is a direct effect of poor glandular function. Meat out of a bull has hormonal content and taste. Indeed, there seems to be no real reason for the practice of castration except to eliminate the bellowing and pawing and behavior associated with bulls. It is that simple.

Feedlots use hormone supplements to restore a measure of hormonal activity in the growing steer. These hormones are essential if the wished for growth is to be achieved. They are also synthetics. The ear tag does not impart taste. It has about the same taste it would have if you ate it direct. Faltering hormonal activity, it seems, is further exacerbated by substituting corn for grass.

Beyond Glands

All the measurements and photographs contained herein only hint at what is inside the animal. For this reason I elect to make a few suggestions based on mature literature and what qualified veterinarians tell me. I do this to key what I think is an appropriate question.

How is the endocrine system altered? How do the hypothalamus, pituitary and adrenal system get out of whack? The hormones are manufactured by glands. The glands take in at one end a molecule called the sterol. They alter it slightly and seed it back into the blood as a hormone, starting with a sterol. The mycoplasma, a mere particle of DNA — usually harmless — if subjected to extreme stress or trauma will start up mycoplasma deletion. In human medicine it is known that the mycoplasma initially enters a cell, lies dormant, then is goaded to action by trauma. It then comes to life. It starts to take apart the cell it is in sterol by sterol. When it kills the last, it does better things. This is called degeneration.

In the cell, plant, animal or human, there are chromosomes which carry almost all of the information needed to direct that cell's growth, division and production of chemicals such as proteins. These chromosomes are composed of information-bearing genes. Radiomimetic chemicals (chemicals that ape the character of radiation), radiation itself, and many of the chemicals used in agriculture can injure the chromosomes either by altering the chemistry of a single gene so that the gene conveys improper information (called *point mutation),* or by actually breaking the chromosome (called *deletion*). The cell may be killed, or it may continue to live, sometimes reproducing the induced error. Some types of cell damage cause genetic misinformation that leads to uncontrolled cellular growth — cancer.

Entry to this mycoplasma is usually denied by a genetic predisposition to keep it out. The signs exhibited by glands collapse when they cannot function at a maximum level.

The picture is even more complicated. The mycoplasma has an appetite for red blood cells. Red blood cells in turn are a sponge for cholesterol, which delivers flexibility. A red blood cell can slide through a capillary half its size because it is flexible. When the mycoplasma removes the cholesterol from the membrane of the red blood cell, flexibility is denied as is travel through the capillaries. The cut off of oxygen and blood to parts of the several systems can be impaired. The input on scrotal function makes it impossible for the bull to express himself as related in earlier passages of this book.

This business of mycoplasma disturbance interfering with the creation of new blood cells and consuming existing red blood cells asks how much blood defect stress and trauma account for. In the human being the shortfall can run from 10 to 25 percent, this according to blood valve tests. I suspect that deformed scrota and misplaced male nipples in proximity to the upper reaches of the bull's scrotum have their genesis in the system industrial agriculture imposes on cattle.

It has been noted in human medicine that serious trauma — gunshot wound, injury suffered in a car wreck, even some vaccinations — often result in the most serious degenerative diseases after some years of apparent recovery. I mean Alzheimer's, Lou Gehrig's disease (ALS), Parkinson's, and so on. Lesser trauma and stress, I believe, tampers with the fine-tuned mechanism of the bovine's physiology. Some parts of this scream from signs and symptoms, some do not. More often the hint for abnormalities is given an arrest by researchers pushing the envelope beyond nature's tolerance.

Freemartins

Freemartins are a good horrible example. The hormone of the male is released first. In other words, testosterone gets into the uterus, then effects the development of the female reproductive system. A complete reproductive system with minor elements missing is often the consequence. It is possible then to have male/female twin calves. Most of the time the female will be sterile and is called a freemartin. The business of freemartins having a sexual preference for their own sex is usually the consequence of the ovary becoming septic. A cystic ovary after a while produces the male testosterone. Such an animal becomes the aggressor in riding females in the paddock. This condition continued long enough permits estrogen to metabolize into testosterone. That cow will take on male characteristics. A similar result is often achieved by researchers who make the cow superovulate with hormone support from dairying innovation pushing the envelope. This procedure ruins that cow.

Withal, the freemartin generally takes on the male characteristics in comparison to the fertile female because she simply does not have the hormones that would preserve for her the status of a feminine cow as measured and observed under the auspices of husbandry discussed here. Freemartins are more steer-like in appearance than their fertile counterparts and often they are more muscular.

It is axiomatic that a freemartin has a part of her reproductive tract missing. If that segment is not missing, she can have a calf. It does not happen very often. The male of a freemartin can service and settle a cow. However, about 40 percent of such bulls exhibit definite infertility.

Cause is too elusive to merit speculation, except to say that faltering genetics can be erased. This can be achieved by ruthless culling and multiple trait selection as suggested by this book.

Growth Plates

It can now be accepted as a given that a bigger cow at maturity means a lower fertility cow. When a bull reaches maturity, his testosterone closes down his growth plates. When a cow reaches maturity her estrogen also closes her growth plates. She devotes her energy from growth to reproduction. Therefore, the highest fertility animals and the longest testing fertility animals are not likely to be humongous in size. The 3,000-pound bull is therefore a deviant, the normal bull being 1,600 to 2,300 pounds. The same is true of the female. Cows that have a highly fertile makeup when young will cycle early and stay fertile for up to 18 years. The widely quoted statement that most dairy animals are good for only an average of 1.5 gestations illustrates general degeneration of both husbandry and species.

To repeat, testosterone in the bull and estrogen in the cow close the growth plates. These plates are in the living bovine, which is why man must measure.

A referencc to the human animal must be in order. A highly fertile person matures early, as evidenced by youthful wrestlers. Such athletes will not grow to be seven-foot basketball players. The growth plates decrease differently. In the human being and in the animal, the ambition for height closes down. The seven-foot basketball player is regulated by late closing of growth plates. His maturity is generally late.

An 800 to 900 pound cow will reach her sexual maturity, will close down growth plates early and use her energy to reproduce. If allowed, she will have a long reproductive life. A cow that gets to be 2,000 pounds will have slow reproductive development. Her continuing growth will slow cycling when young. As size at maturity nears there is heavy milk production — all at the expense of fertility and long life.

Additionally, the stress factor for cows that are on concrete cannot be calculated, only imagined. All these things conspire to reduce gestation to the statistical average of 1.5 per animal.

Fertility and Twins

My associate, Charles Walters, has noted that when a breed has fertility problems, a much higher incidence of twins is a consequence. Holsteins have lots of twins. Respect for the knowledge of seasoned cowmen suggests discrimination against twins. Twinning is not only hard on the cow, it also represents departure from fertility. I know researchers are working on the idea of twinning, but it is stupid. A cow that twins calls for maximum attention, almost hospital attention. Twins to a dairy cow is usually the kiss of death in the concentration camp environment now fostered on dairy and beef. The twinning gene should be considered a harbinger of infertility.

I have in mind one research project that involved 150, 150 and 150. They took single heifers and bred them to beef bulls. They got about three percent twins. They then took twin heifers and bred 150 to single beef bulls. They ended up with 19 percent twins, a dividend from their mothers. They then took the next 150 twin heifers and bred them to twin bull calves. The result was about 38 percent twins. The bull now affected the count.

When the Simmental came in, the closer it was bred to a pure Simmental, the greater the twin issue. The Simmental is a very productive animal in terms of milk and growth, but it is a high-maintenance animal. The correlation seems clear enough. Infertility correlates with twinning. Twinning in turn correlates with a high metabolic rate.

Trauma, stress and genetics are debilitating knives that cut several ways. Cowman Jim Lents of Indianola, Oklahoma, put it this way, "the industry has turned the bovine herbivore into a ruminant hog." Pasture defends *Bos taurus* and *Bos indicus* from that fate.

Chapter 6

Pastures

Almost a century ago, in the New Jersey laboratory of Dr. George H. Earp-Thomas, it was proved that the forage called grass took up micronutrients in sizes beyond the ion or angstrom level. It was believed then, and I believe now, that herbivores function best on grass. The attempt to substitute corn for forage, with animals confined and force-fed on a high carbohydrate diet, has enjoyed the blessings of academia but there are serious flaws in this substitution. Even the switch from open-pollinated to hybrid corn is fraught with danger.

Corn is an errant member of the grass family. Some of the successes that have been seen with corn were probably the result of distorted accounts. Still, in terms of nutrients, open-pollinated corn has an enviable record. Some 4,000 samples of corn were taken from ten Midwestern states in a single year. Open-pollinated corn (also called OP corn) contained 75 percent more available protein, 87.5 percent more copper, 34.5 percent more iron, and 20.5 percent more manganese. The same general trend has been identified for calcium, magnesium, sodium and zinc. It can therefore be said that OP corn contained an average of 400 percent more base nutrients. The failure of hybrids to uptake certain nutrients was confirmed in the laboratory of the Armour Research Foundation in Chicago. The comparison was between Krug OP corn and a regular hybrid. Spectrographic testing revealed the hybrid short of nine minerals. The hybrid failed to pick up copper, and any other trace minerals. Both varieties had the same chance to pick up a balanced ration. The failure to pick up nutrients is what had implications.

The core of vitamin B-12 is cobalt. Ira Allison, M.D., and others have found that a lack of cobalt was implicated as a cause of brucellosis (Bang's disease) and its human variant, undulant fever. A lack of cobalt is the cause and its supply is the cure. Hybrid corn makes poor farming by producing bins and bushels without substance. Yet this is the chief feedlot nutrient, a substitute for forage.

A newer corn styled Bt worsens the situation because it produces a red *Fusarium* mold. This mold debilitates and kills in swine operations. It also has been indicted for canceling out the ability of males to breed and sows to conceive.

Cowboy arithmetic does not start or stop with genetic selection or the fundamentals of management related in the last chapter. It asks us to examine the pasture in terms of the bovine — cow, bull or steer. For the purpose of these passages, cow means bovine and grass means forage.

The cow is truly one of nature's masterpieces. Rudolf Steiner, the master who arrived when a select group of farmers were ready, used Einsteinian insight and saw the cow's four stomachs as a laboratory par excellence, with the rumen as a fermentation vat, and traffic through the reticulum, omasum and abomasum as the finest distillation of bacterial strength. In addition to the cow, sheep and goats rated attention as common farm ruminants with similar capabilities, but the cow was queen. The anatomical diagram below suggests a processing apparatus in which bacteria attack food as follows:

Several hundred species of bacteria have been identified in the rumens of sheep and cattle. Guinea pigs and rabbits are herbivorous animals, but they do not have a population of bacteria

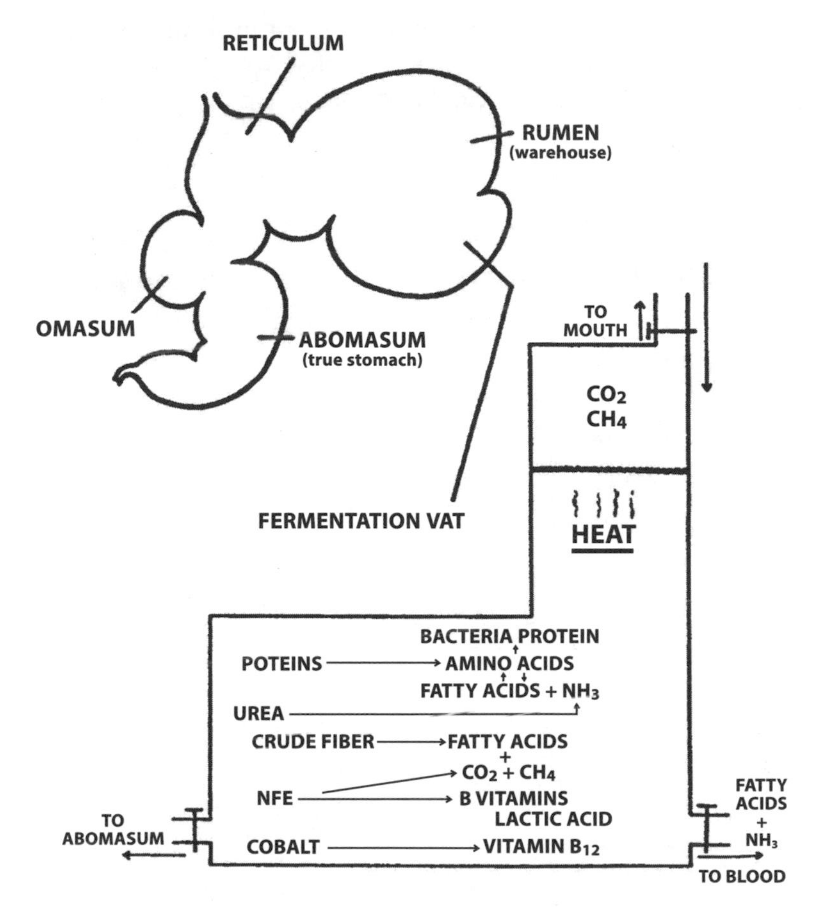

The laboratory that is the four stomachs of the cow.

in the foregut sufficient to accommodate maximum fermentation of ingested carbohydrates. Only the cow achieves climax perfection of acid, air, water, heat and mixing conditions.

Nature has decreed that forage is absolutely vital to this animal's health. Harvested and preserved feeds never equal the star quality of green forage, that superb pasture mix of 50 or more species farmers simply call grass.

Indeed, John J. Ingalls called grass "nature's benediction," and scientists have confirmed his assessment by identifying vitamins and minerals that only grass seems to provide. Many, if not most, vitamins are rapidly degraded in the harvesting process.

Vitamin A, for instance, is abundantly present in fresh grass. Hay that is dried and the grass called corn that is made into silage when preserved, all lose or badly decompose their vitamin A content.

Self-Evident Observation

Grazers find this observation self-evident. They observe a paucity of disease problems when cows are kept on grass. Cowmen who use rotational grazing as a management tool tell me that animals are advantaged health-wise in several ways. Grass supplies mineral nutrients in a size that easily serves the digestive process. The animal seems to have both the nutritional knowledge and the discipline to adjust its intake volume. Scientists have pretty much settled the fact that on adequate pasture, the bovine consumes about three percent of its body weight every day, this on a dry matter basis. This arithmetic suffices for a lactating or growing animal.

The grazing animal moves around in order to gather in its daily diet. Its tongue selects a tuft of grass and mows it down with teeth situated in the lower jaw. There are no upper teeth. The grazing animal moves; therefore, it is always a bit athletic and in better condition than a feedlot counterpart. Feet and legs are stronger and healthier. In other words, pastured animals last longer.

The pasture basked in fresh air contrasts sharply with the iron-railed cowpen forever clouded with an aerosol of fecal dust and toxic sprays. On a grassland sward, the cow inhales vitamin D from the sun. Not least, the pastured animal has a clean bed at night. Withal, it is the cow's digestive system that must be considered when pastured animals are juxtaposed to high energy feedlot victims.

It is a 45-gallon fermentation tank that best distinguishes the bovine. That tank turns grass into metabolically useful organic acids and proteins. The "four-stomach" system is a superb key to nature's balance, always a correct link between animal and human being. The uncommon good ecological sense of grass and rotational grazing will be unfolded soon enough. For now, I would like to ask and answer questions feedlot afficionados avoid with the armor of ignorance.

There is an in-depth story behind the high-tech paradigm that takes a cow off grass during the terminal months of its life, but it can be summed up with one word: speed. Admittedly, cows grown on grass take longer. Cows force fed on a rich diet of corn, antibiotics, hormones, protein bypass, etc., fatten at an indecent tempo, producing a product so tasteless it is rapidly eroding consumer acceptance.

Four Years

There was a time, circa World War II and earlier, when it took four years to finish out a steer. By the early 1950s, the time allotted the average beef was two to three years. The university took the lead and feedlots were on the way. Nowadays, 14 to 16 months sees the calf's life terminated and a carcass on the rail.

Taking a calf from 80 to 1,200 pounds in 14 months cannot be accomplished with a magic wand. It takes corn, protein supplements, growth hormones and drugs. The new tendency has funneled off profits from the primary producers by installing a high-volume and low-profit scenario into the countryside. The mere idea of finishing beef cows or maintaining dairy animals on pasture has become anathema to the "conventional" farmer.

There are cowmen, now in their nineties, who swear they made more money with half as many cows in the 1950s as their sons now make with two and three times more animals using chemical-hormone technology demanded by the industrial model.

Recasting a biological procedure into an industrial procedure starts with weaning on the industrial farm. Those who prepare their animals for the long ride to the feedlot sometimes wean calves and start high-energy feeding cold turkey. The stress is awesome. The super-rich diet ushers in illness. The quick post-weaning trip to the sale barn is also attended by a great deal of sickness and shots, a veterinary intervention, and sale barn docking for this and that.

Corn, alfalfa hay and Rumensin — the corn and Rumensin ration ever increasing as alfalfa hay decreases — assault that digestive system much as might any Frankenstein food.

Rumensin, Tylosin, corn and corn silage, liquefied fats and proteins formulated nowadays, if not previous generations of dogs, cats, cows and other animals, a diet now prohibited by USDA — synthetic vitamins, estrogen, antibiotics, molasses and urea — these have become the substitute for pasture.

There are problems with these substitutes for grass. Corn feed means more saturated fats. Pasture-fed meat has a great deal less fat than grain-fed meat, fats in grass-fed meat being healthier. Grass-fed meat has more omega-3 fatty acids and fewer omega-6, the author of heart disease. Fatter meal, pig and fish protein, chicken manure, all are approved cow feed. Rendered animals, of course, are fed to chickens, and chickens and their droppings are fed to cows, albeit not without nature's disapproval.

Health Problems

On the farm and in the feedlot, health problems trace back to diet. Cows are herbivores and this means they are designed to consume forage. Birds are designed to eat grain. A ruminant on grain is subject to bloat. When a diet of starch closes down rumination, gas becomes trapped in the rumen. Gas presses on the lungs. Suffocation follows if a tube introduced into the rumen doesn't relieve pressure.

Corn also creates acidosis. A natural pH in the rumen is 7, or neutral. Corn and feedlot fare often bring on acidosis so severe it kills the animal. Animals affected by acidosis go off their feed, pant, salivate, paw their bellies and eat what passes for soil in the pens. Diarrhea is a sign. Ulcer, bloat and liver disease trail along as the animal is led to slaughter. If feedlot pneumonia and polio do not intervene, six months on feedlot fare is about as long as a critter can make it. The suggestion that such meat is healthy is astounding, indeed.

Pasture O' Plenty

A pasture does not supply plenty simply because it is green. Availability of micronutrients depends on the quality of soil, rainfall, how well the natural carbon and nitrogen cycles are working, the management of paddocks and the level of stocking. The idea that a pasture is complete and the cow a miracle worker capable of transmuting deficient grass into micronutrients is an economic death trap.

There are geographical areas that are seriously copper deficient, southern Ohio, for instance. Pasturing cows in that area calls for mineral supplements as a given. Copper is not the only micronutrient that might require special attention.

The basic rations have to be considered as an opener — calcium loading the soil colloid to the tune of 65 percent and magnesium commanding 15 percent of the colloidal positions. The basics of the row crop soil manual plug in, of course, and Andre Voisin's *Soil, Grass and Cancer* should achieve biblical attention.

The bovine species is in a class all its own. Sheep do not need as much copper simply because that species does not have as high a copper requirement as the bovine. I note this because there appear to be differences among cattle. According to work done at the University of Kentucky, Jerseys do not have the copper requirement of Holsteins on the basis of body weight.

Pasture Management

Pastures and management intensive go together like ham and eggs. David Zartman of Ohio State University put it this way in an *Acres U.S.A.* interview: "What you are doing is taking yourself off the tractor seat and going out on your feet, turning your head on instead of your tractor's engine."

Grass farmers have to be management intensive pasture people. They have to nominate the bovine animal as the harvesting machine. That animal also distributes fertility and constructs a nutrient-balanced system. Inputs tend to balance outflow in the fullness of time, meaning meat, milk and eggs going out, inputs meaning mineral, forage, sunshine, water and even nutrients transported by air coming in. The rhythm of balance becomes a fixture within a few years after management intensive procedures are installed. Animals consume the nutrients and recycle them. They harvest and fertilize. They recycle every 14 to 16 days early in the spring and fall. They do the same every 30 days in the summer months. Irrigation of a strategic reserve pasture for summer survival may be an option, this based on an inch of water a week used on golf courses to cancel out early summer dormancy.

There are a few rules that cannot be ignored with impunity. First, the pasture has to be of reasonable composition and density. About 400 pounds of dry matter per edible inch on a per acre basis is a reasonable expectation. This inventory is defined by the rotational sequence. A 14-day rotation cycles back to the starting point in two weeks. With, say, eight inches of forage, cows will eat five or six inches. That means five inches of 400 pounds dry matter per inch, one-acre basis, in 14 days.

Divide the acres by 14 to determine the forage available each day. Acres plus yield per acre equals the volume of forage in pounds per day. Divide the above by 0.03 (three percent) to compute the pounds of forage animals the pasture can feed.

Obviously, the cow demands more forage than a goat herd, and a lactating or growing bovine animal demands more feed than a mature non-lactating or non-growing animal, the latter requiring only two percent of their body weight in forage each day. Body mass governs.

There is an old saying that all absolutes are false, including this one. No plan is perfect. Hard headed stick-to-it-iveness is no virtue. The natural plan may falter, even fail. The pasture usually delivers more forage than measure and computation indicated. If a rotation does not comply with expectations, then the rotation delivers a new problem — understanding. A rotation that does not comply with those great expectations tends to deliver overly mature pastures at the rotation's end. Resultant haying problems and rotational pasture out of sync become the consequence. The answer is flexibility and routine monitoring of paddocks, one at a time and all of them together.

The cowman's eye for body condition now comes into play. There probably is nothing on earth as contented as a contented cow. Cows will communicate when things are right and when they are wrong. The cows, not the predetermined system, should call the shots.

Climate, weather, and not least, rainfall, alter the rotational system faster than a draftsman can erase a line on a blueprint. Very wet weather can mean runaway grass. A dry spell can order fescue to go to seed. It takes a routine monitoring to know when to clip, seed heads in the pasture being a no-no.

Moving animals more frequently and changing the numbers is a function of management, always with body weight computation and feed availability in mind — feed divided by 0.03 body weight being the formula. If enough pasture is not available another feed source, supplementation, population reduction — in a word, maturity of animals — all contain some parts of the answer. Turning an inventory of cattle into cash may make more sense than loading the operation with unrecoverable expenses. Withal, grass, not ADM-touted soybean fields, holds in escrow the key to acre profitability.

The University of Illinois has an excellent set of data on small grains versus pasture in terms of bottom line. Data reveal that small-grain acres deliver a net profit per acre of no more than $80. Without the government's deficiency payment program, the bottom line number would be negative. The management intensive pasture system developed a $600 per acre net profit. The numbers for fat cattle or nondairy show a $300 per acre bottom line as a result. Some dairymen reported more than $600 an acre, some over $1,000 an acre. Most of the above were seasonal dairies, not year-round operations.

Absence of a pasture program tends to mean high maintenance. Farmers who learn to embrace nature's rhythms discover grazing as a sort of Holy Grail. They accept the requirements of being seasonal. Seasonal maximizes net income. Fall freshening offers results.

The Grain Connection

Between 70 and 90 percent of the grain grown in North America feeds livestock. That's what the parity fight of the 1950s was all about — to reduce grain prices to a world level. This, in turn, animated massive cow pens as high tech and downplayed pasture as "horse and buggy" animal husbandry.

That is why grain farmers get a "deficiency" payment to keep themselves and feedlots in business. Low-priced gains on the market in effect subsidize cattle pen feeding described earlier in this chapter. Globally the grain fed through livestock is only 40 percent. Yet herbivores, even omnivores, can get their entire diet from forage. Even chickens can replace a part of their grain diet with grass and insects.

Opening the soil to erosion and degradation is not the only shortfall associated with maximum grain production. There are the pesticides, herbicides, energy costs, and NPK salt fertilizers, all of which invite the twin engines of erosion — wind and water. All could be greatly reduced with pasture.

Allen Henning's advice now comes to the fore. Henning is a grazing consultant who spent some 15 years in New Zealand. When farmers in the Midwest asked him how to convert row crop acres to pasture, and asked what should they plant, he responded, "Nothing at all." This is not what most farmers want to hear. Henning tells farmers to take old pastureland, cut it in halves, one half to be grazed rotationally, the second half to go to seed. Graze for three or four hours on the good part, then a similar amount of time on the seedy part. The rest of the day is spent on the old worn-out field. The nutrients are distributed with the manure. Over time the pasture establishes itself on row crop land.

Admittedly, this takes time. But there are other examples. Ann Clark of the University at Guelph, Ontario, tells of a New York grazer who bases his income on abandoned land. He finds the owner, gets permission to fence it with five-strand, high-tensile wire, then brush hogs it, then cross-fences for rotational grazing. He uses no seed, no fertilizer, no nothing.

Always, there is a dormant seed bank in the soil. There are limitless grass and weed seeds that stay on for years, decades, even centuries, awaiting permission for life. When trees are removed with a brush hog, permission is granted.

As a practical matter, conversion of grain land to pasture should be accomplished as cheaply as possible. Broadcast seeding with animals trampling seeds into the soil, land to be divided into perhaps four paddocks is sufficient for a start.

There can be no manual. Each farm has its own geography, environment, water flow, natural entry and egress, gate location, paddock divisions. The experts are often the neighbors or associates with experience. Clark advises farmers to go simple, to put out as little money as possible with most of the money going into permanent fences. Interior fences can be simple.

The problem with seeding is that what works in Wisconsin does not work in Mississippi, Alabama or New Mexico. In the main, complex mixtures work best, five or six species carefully chosen with matched maturity and palatability and one or two legumes probably figure. White clover may figure in Canada, whereas red clover is more likely candidate for Louisiana, Arizona or Colorado. Some kind of legume is critical. A good bottom grass is a prime requirement for the mixture. Cooler climates like Kentucky bluegrass. Rain belt acres might consider a little perennial rye grass, two or three pounds per acre. It establishes itself fairly fast, structuring a soft floor quickly. It takes a good sod to hold up the animal's weight.

If land has been out of grass for ten or 20 years, a legume is mandatory. It has a capacity for recovering the natural nitrogen cycle. Its seeds have a dormancy equaled by few, which explains why a return to forage is almost always announced by the arrival of clover, always without one seed being planted. The grasses in the mix should be local grasses, the kind any extension agent might recommend, the nature of spring, fall, summer and winter being considered. The level of experience must guide, and this principle must be invoked to protect the amateur and dilettante. Different mixtures for different parts of the farm also are indicated. Every farm has high acres and low land, north facing, south facing. Each situation invites special study and answers.

A natural summer pasture, an insurance pasture if you will, needs deep-rooting species. These will not exhibit strength in the spring, but they will deliver when other pastures have become exhausted by sun, heat and wind. The wrap-up is simple: complex mixtures, one or two legumes, a bottom species, and then species indigenous to the location.

A Pasture Mystique

All the above notwithstanding, consensus thinking among dairymen seems to be that there is no place for pasture in the milk production business. At great conferences the tales seem to embrace the idea of taking acyclic cows and turning them into cyclic cows with various injections and then extreme measures.

Ann Clark put it this way, "These [speakers] are all men. They've never given birth. They don't know that when a cow doesn't cycle it means something. She is under stress. And rather than just back off on the stress and reduce the conflict between lactation and gestation, they want to force her to both lactate and cycle."

The pursuit of unnatural systems of feeding is excused on the premise that grass as the chief feed ingredient inhibits size. Admittedly, the Horizon-type organic farm — truly an oxymoron — would be impossible as a strictly grazing operation. Yet there are a number of farmers in New Zealand that keep in excess of 500 head of cows on grass. There is one dairy in Wisconsin, according to Ohio's David Zartman, that has over 400 grazing dairy cows. The former Ohio state dairy specialist, now a prime mover in pasture management, suggests 1,000 animals on grass as the top level, depending on how many milking parlors the farm wants to field. It is not practical to walk the cows a mile to the milking parlor. A half-mile should set the limit.

Lessons for cow maintenance are certain to include pasture, and management quite literally draws pasture management lessons from New Zealand and Ireland.

In a manner of speaking, the intellectual advisors to the farmer set up several paradigms in the 1950s. The dairy animal led the way, and the beef cow and steer followed. First, the more milk a cow could be goaded into giving, the better. Second, anything that could be purchased to achieve production ought to be purchased. Third, each cow should be managed as an individual in order to optimize input as related to output. Fourth, it was concluded that manure was waste and therefore a disposal, not a utilization problem. Theoretical principles were implanted so thoroughly, cowmen — first dairy, then beef — found it rejection of their heritage to let go of them. Pasture was downgraded and rejected in dairy, and beyond cow-calf operations in beef.

As a consequence, input costs in dairy have quadrupled.

Thus few operations leave open the door to the future of pasture, pastured beef, and improved genetics — in short, a cut of meat that supplies a satisfying dining experience. There is afloat the opinion that concentrated feeding days are numbered. The 1995 Farm Bill failed in its attempt to dissolve grain subsidies. But the congressional impulse is strong to have done with maintenance of wealthy landowners in order to market grain at world prices. World-priced grain is absolutely essential for continuation of feedlots. ADM's oft-repeated TV ad about turning pastures into soybean acres suggests the desperation of the feedlot system to maintain the status quo. At an average feedlot profit of three dollars per steer, the equation is getting mighty thin, and the pasture potential is reasserting itself.

Added to the above is the intensity of the environmental movement, and the quite definite charge of air and water quality destruction. The EPA wants all water to be fishable and swimmable, a mandate itself at odds with feedlot environments and perfectly at home with pasture.

The massive feedlot — after harvesting most of the wealth created by the cow-calf farm — is a sunset operation. In addition to the two truths stated here, there now asserts itself the animal rights movement. Humane treatment of animals is the issue. Consumer pressure is the weapon. It is no accident that chains such as Hen House in Kansas City now feature grass fed beef, pastured pork and free-range chickens.

The several factors conspire to erode that three-dollar-per-animal profit still supporting cowpen operators. Three generations ago, beef producers made more money with pasture than they now make fattening and backgrounding calves for feedlot finishing. The goal is to return to that plan. Genetics and pasture stand ready to make good the right to receive payment for as much wealth as cow-calf farmers have created.

Growing Grass

To speak of the discovery of grass would seem absurd unless such a statement was modified to add, "how grass works." This discovery took place at the beginning of the last century in the New Jersey laboratory of G.H. Earp-Thomas, working under the auspice of the Rockefeller Institute. The Institute did not release the summary report for reasons never declared, but Earp-Thomas reclaimed his role almost a half-century later after his laboratory was torched under unanswered circumstances. Earp-Thomas' study of how green plants grow ranks as the greatest step forward in growing in 500 years. Simply stated, Earp-Thomas grew plants in a special jelly that permitted identification of nutrients and microorganisms. This enabled scrutiny of root hairs feeding, microorganism entering and leaving the plant, and the transport of nutrients. Under magnification it was revealed that microorganisms actually passed into the plant structures. As beneficial bacteria were introduced to the special jellies, a veritable war of the miniatures could be observed. Certain species could multiply rapidly and annihilate the beneficials. Thrifty plants could be laid low when platoons of the infection and toxins were created by invaders. In the laboratory, it was up to the scientist to make the right jelly. In the paddock, it is up to nature and nature's helpers to assist the environment.

In the laboratory the worker strives for a sterile atmosphere, but this is synthetic reasoning. The water pasture asks for beneficials to dominate. Inoculation of row crop seeds became state of the art not much after the process was ratified by a Rhode Island station. This left unresolved the question, How does grass achieve its primitive perfection as a grass feed for the bovine?

I can only make a suggestion, one encountered on a Minnesota ranch. I offer this as a vignette and as a plea to see what you look at.

Cowography would be a term in need of invention except for the Elko, Nevada, cowboy poetry gathering each year. In a real sense, cowographers have been around a long time. They see what they look at. Wesley Ervasti is a legendary cowman proprietor of the Bow Tie Ranch near Sebeka, Minnesota.

Over his many years in farming, Ervasti noted those bare, wet spots on a cow's nostrils. A healthy animal always has what appears to be beads of sweat. Except for Wesley Ervasti, no one has been curious enough to ask the question, "What is that exudate?" Professors and experts have had no answer.

Ervasti figured it was the most wonderful culture ever devised by nature. When a cow mows off grass or eats hay from a bail, she cannot avoid leaving some of the culture behind. This culture helps the animal with digestion.

Mother Nature made a deal with the cow. She said, "I'll provide you plenty of grass, but you have to give something back. Every bite you take, you will leave this culture in the forage and in the soil." Nature did not intend that there should be a piece of soil tilled only by old iron, the plow or cultivator. Such instruments do not bestow the culture a cow provides via her bartered exchange for grass.

Ervasti used the above information on the ranch he bought at age 17. The very same culture, he found, was evident in buttermilk. The same culture helps a cow with her own digestion, and it helps the soil keep its life-factor.

A farmer's pragmatic observations do not achieve "scientific" status when scientists ignore nature's hints. There are no present answers as to what this bovine culture does for the soil. Ervasti has his suggestions, just the same.

He bought the original 160 acres from his folks at age 17. As neighboring farms came up for sale, they were rundown in fertility, and the sale price was almost always cheap. Ervasti made his acquisitions, fenced off the plots into 40 acres each, and returned row-crop acres to pasture. After grazing the weeds and forage for a couple of years, and after manuring the ground, the rejuvenated soil was returned to rotation.

"The cows put on their culture for me," the cowman said, "and they seeded the soil with their manure."

His rotation has been corn, a nurse crop of oats, alfalfa and grass, with hay claiming the land for two years. Cattle would graze the next two years for maximum culture distribution. Corn would follow.

Yields were always fantastic. Although Ervasti has shared his findings with cattlemen, farmers and professors, few have embraced nature's mystery with the humble interest of this one farmer.

He made his suggestion to Dr. Jim Brinks of Colorado State University, an animal science specialist. Brinks agreed that there had to be a reason for that culture on a cow's nose. He turned the idea over to a botany specialist who harvested this culture, diluted it with water, and sprayed it on potted plants. The usual treated and untreated plants anointed the greenhouse. There was a 50 percent increase in those treated with culture.

Dung Beetles

One man's experience with a perceived cow culture merely opens the box on pasture lore, much of which is ignored or not understood in the first place. Truman Fincher, while a researcher at College Station, Texas, computed that about one billion cow pies hit the pastures each day. The loss of grass and nitrogen, and the loss of production due to parasites dung beetles would control, cost owners about two billion annually. How to dispose of manure in a manner calculated to preserve the economic value of each dropping became a task of passion on the Walter Davis ranch near the Red River on the Texas-Oklahoma border. Davis has about 3,000 acres of pasture and uses planned grazing. Working with Fincher, he populated his acres with dung beetles. Nature's own sanitary engineers keep his pastures clear of dung within hours during the warm months of the year. "Ideally," says Truman Fincher, an area needs eight to ten species of beetles to police an area, these to exhibit a minimum of direct cooperation. This means day fliers, night fliers, rollers, tumblers, and dwellers.

Unfortunately, USDA has closed down the laboratory facility that propagated, colonized and, in general, supported a drive to undo the damage accounted for by pesticides. Cowmen who seek the aid of nature's own sanitary engineers — tumblers, rollers, dwellers — are on their own, as is the case with all innovators.

The business of taking cow dung underground within hours suggests an efficiency that is denied grazers now that man-made chemicals have upset nature's rhythms.

Bacteria, fungi, mycorrhizae, all confer humus to soil, the prime requirement for rich, abundant forage. Pat Behrens of the University of Texas called them "organic gardeners." Dung beetles love dung. They feed on it, rear their young in the nutrient-rich medium and, more important, see to it that the choice contribution to soil fertility is not wasted. Their home is down under, and the platter load goes with them, if they are allowed to propagate. They have no teeth, no bite. They slurp the available fluids.

The economic benefits of any successful release are at once apparent.

1. Rapid removal of cow and bull droppings prevents valu able grass from being smothered or made unpalatable.
2. Rapid incorporation of dung into the soil prevents loss of nitrogen.
3. Dung beetles short-circuit the life cycles of many insects and internal gastrointestinal parasites.
4. A healthy beetle population removes breeding media for horn flies and fireflies.

A Biodynamic Approach

There are dozens of pasture sprays now a part of the eco-farming scene, most of them quite commercial, all of them effective to the degree claimed. I cannot detail the range or content, yet I would like to relate an experience I had in New Zealand. Making the different ranches (stations in Australia), I encountered a product called horn-manure, more specifically, The Prep. Mere recital of this art form is insufficient. To follow the art requires study, a whole new look at the literature of Rudolf Steiner, Alex Podolinsky, Hugh Lovel and Ehrenfried Pfeiffer. Suffice it to say that the basic prep is made by inserting fresh cow manure from lactating cows into horns reclaimed from the abattoir. Filled horns are buried over the winter months, equinox to equinox, then recovered in spring. By then the horn manure has become charged, transmuted. It is then cleaned and ready for application to the pasture as a spray. The spray is created by putting a fix of horn manure into water, stirring it for about a minute, or until a vortex has been achieved, then counterclockwise to the same effect, back and forth. The theory is that the organization of the resultant spray benefits from "an evolution of organization."

I was dumbfounded to see grown men burying hundreds, even thousands, of horns in order to grow a micro mix for pasture spray of three gallons per acre. Stirring can be done by hand, but in fact most ranchers down under use machines and create drums of mix with each mechanized hour-long stirring.

The spray is applied at dusk. For maximum effect, a second spray is applied the following morning. This one is called the Horn Silica Prep, created by a summer-long residence in the soil.

The evening treatment calls for heavy droplets that sink into the soil. The silica prep the following morning must be a fine mist that aerosols into the atmosphere. Followers of Steiner and Podolinsky are often at odds about fine points of detail, but the fact is that this pasture treatment is widespread. My associate, Chuck Walters, the long-time publisher of *Acres U.S.A.,* visited many farms in Australia, Tasmania, New Zealand, and stateside. At the time he visited New South Wales, Alex Podolinsky was expanding the Biodynamic following exponentially always without a nod from the republics of higher education.

The sometimes pontifical pronouncement has it that use of the evening Prep lasts a few days, but followed in tandem with the Horn Silica Prep, the effect lasts much longer.

Horn manure improves activity in the soil, including billions of unpaid microbial workers, bacteria, fungi, mycorrhizae, etc. The stir of nutrition for the grass can be imagined. It helps fill out the grass with a nutrient load. All atmospheric activities, namely photosynthesis, blossoming, fruiting, ripening, all are addressed by the Horn Silica Prep.

My exposure to the Biodynamic art is so new, I hesitated to add it to this chapter on pastures. But thousands — even a million acres — is too much to be ignored.

Still cycle students tell me that sections of the U.S. — especially the Texas panhandle — can expect a 20-year drought and the hazards that go with a shower flush and rapid growth of drought-stressed crops. I am quite mindful of the fact that drought-breaking rain can spur growth, sometimes releasing toxins, prussic acid, and a derivative of cyanide.

Prussic acid poisoning can follow. This is more likely when cows are turned into milo fields after a drought broken by a shower, but Johnson grass also can deliver the prussic acid haymaker. Release of nitrate toxicity under the drought-shower circumstance is legend, for which reason some farmers never, never buy nitrogen as in N, P and K, preferring to work the nitrogen cycle instead.

I mention all this because the New Zealand and Australian grazers claim exemption from many of nature's penalties when they use their horns of plenty.

I will neither recommend nor discourage the use of this approach, except to encourage those who are interested in it to study the art. Preps are prepared in the United States by the Josephine Porter Institute in Virginia. This organization has been a guiding light for years. Podolinsky's lectures are generally available on eco-farming bookshelves, as is Lovel's *A Biodynamic Farm,* and all are available from Acres U.S.A. in Austin, Texas.

Plants in Action

Plants trap solar energy in the form of simple sugars. Stated simply, the reaction is $12\ H_2O + 6CO_2 \rightarrow C_6H_{12}O_6 + 6O_2 + 6H_2O$, mediated by the presence of chlorophyll. This chlorophyll is the catalyst responsible for the green in plant leaves and grass. Photosynthesis is the basis for almost all life. This eat and get eaten routine is the basis of the food chain.

This brings us up to the carbon connection. Plants regulate carbon by taking in and converting carbon dioxide. This is the way carbon, essential for organic molecules, can get into the cycle we call life. This staff of life is really extracted from the air envelope surrounding our planet.

This is why plants define and control life.

We can go on and tell how tropical trees control the water cycle to a large degree. Plants are sensitive, not only to rain but also to chemical abuse, the toxicity in rain and to the lack of rain.

One can visualize the carbon sinking role of grass, and the suggestion can be made that only organic acres account for maximum performance. Yet the single cell algae near the surface of the world's oceans crank out more oxygen than the entire inventory of the world's land-dwelling plants. Of course, more than 70 percent of Earth's surface is ocean.

This information is presented as an aside and to identify the premier role of grass in making the life connection. This connection is broken the day an herbivore is reduced to existence in a stall, to be maintained on computerized feeds that make little or no room for the green requirement.

This short aside is a reminder that all plants do not meet the green standard of grass for mineral and vitamin uptake as well as synthesis. With the deficit of hybrids noted, some attention must be directed to Bt corn.

In the April 29, 2002 edition of *Farm Bureau Spokesman,* Bt corn was indicted as an unnatural interdiction in hog breeding success. Bt corn can therefore be suspect when a bull's performance fails, as has been suggested when soy estrogen-laden fillers are fed to the bovine male. Bt corn carries a microscopic red mold generally identified as *fusarium.* Sow farrowing rates have been reported faltering 80 percent under conditions of Bt in the rations. The Farm Bureau report did not cite only a single case report, but multiple farms, all exhibiting the same breeding syndrome.

Red mold added to the trace mineral defect in hybrids simply has to be a factor, even though debilitation can be slower in the bovine. The *fusarium* strain need not detain us. But at least two are quite lethal when concentrated.

They are *Fusarium subglutinans* and *Fusarium monilif.* A switch away from Bt corn terminated pseudo pregnancies in the case of the hogs cited above. The problem of Bt corn may not be erased by a simple rotation once the errant gene has been installed on row crop acres.

Grass has phenylpropanoid compounds not available in grains, protein bypass, cattle cake or any of the fabricated feeds based on manure, previous generations of animals or fish. There are in fact members of the healing arts who extract these compounds via distillation. They capture essential oils by opening the fibers of plant tissues so that molecules can exit with the steam.

These essential phenylpropanoids are sheltered by a sac. Even distillation in its high-temperature phase can't crack that sac. But as cool temperatures take over, the sac is ruptured, delivering the vital material into the cold water. A centrifugal separator does what the milk separator does.

This special property of grass just happens to be almost identical in function and purpose to the blood of the cow. In the animal, blood and lymph fluid invite our attention. The plant's markers are chlorophyl and the phenylpropanoids. Blood feeds body cells. The various oils — according to species — feed the plant.

Blood transports antibodies, antibacterial, antifungal, antiparasitic, and immune-stimulating materials. It just so happens that the same properties are contained in the life fluids of grass. In short, the life fluid of grass and the blood of an animal have almost identical functions.

You've seen carnivores graze when they're sick — the dog taking to grass and forbs when they have a problem. Grass contains natural hemostats.

It is no accident that good pastures have perhaps 40 to 50 species, many disguised as grass. Nor is it an accident that fully half the weeds in the USDA index are also listed in manuals on medicinal plants.

The bovine animal gravitates to species needed for digestion, the call for laxatives. That is why cows seek out wild onions, green shoots for the laxative factor. The late William A. Albrecht often pronounced the cow superior to the best scientist as a nutritionist. They have an instinct for selecting forage for its detoxifying effect. They instinctively select plants for the antibiotic effect. Given suitable forage, they stimulate their immune systems. If you watch cows at grass the way André Voisin did before writing *Soil, Grass and Cancer*, you'll discern the cow as her own veterinarian.

Before a nutritional element can be carried into the cell, it has to arrive via the plant in a size suitable for instant assimilation. That is what separates the mineral box from the blessing of grass. A micron-sized nutrient must first be acted upon by rumen microorganisms, a function ion-rated nutrients are excused from demanding.

Oxygen, of course, is the catalyst. Take oxygen out of an animal's blood, and the result is death. Remove the oxygen from grass, the result is dead fiber.

Grasses, meaning forages, are sources of protein. They are amino-acid compounds.

I think this inventory of information is why John J. Ingalls called grass "the forgiveness of nature, her constant benediction."

The best grass is native grass evolved for a special terrain over eons of time. This is why I counsel against moving cattle from one end of the country to another.

Pioneers who penetrated the continent always moving west did not understand what cartographers call the isohyet line. It runs approximately along the 98th Meridian, from the Rio Grande to Canada. Students of history have to chortle over debates that attended the arrival of new states as the nation was being settled. The isohyet line in effect said that plantation slavery wouldn't compute west of the line.

At the 98th Meridian — give or take a bit — rainfall averages 30 inches. The trees and shrubs at that geological divide are the same species, they have the same DNA as trees and shrubs found along the rainbelt line called the Louisiana border.

For every 15 miles east of the isohyet, there is an additional inch of rainfall per annum, on average. For every 15 miles west of the line, rainfall is diminished an inch per annum, on average. The grasses nature has evolved fit perfectly into their natural terrain.

When Frederick von Rohmer came to Texas in the 1840s, he viewed with wonder the sea of grass that crowded out brush and cedar trees. This was virgin grass, but virgin grass overgrazed is virgin only once. Overgrazing west of the isohyet is several times more lethal than overgrazing east of that geological line.

The Texas Hill Country lies west of the 30-inch rainfall line. When settlers saw the lush, deep-rooted grass of the area, they believed their dreams had come true. They planted cotton. They forgot — as ranchers still forget — that grass grows not over a season, but over centuries. It wouldn't have survived in the Hill Country or the High Plains except for occasional fires, set by lightning and Native Americans. Fires cleared the land of underbrush, the relentless enemy of grass. Brush steals moisture, shuts out grass. Without intervention, brush destroys grass, yet grass outgrows brush. In the wake of a prairie fire, grass gains the upper hand. By the time a post-fire brush plant arrives, grass is thick and strong enough to dominate.

Virgin grass bestows a padding atop the soil, a mulch of sorts that protects the soil, rations moisture, and holds nutrients in escrow. In the scheme of things, the several species of native grasses defended each other.

When the soil beneath grass is thin atop the limestone payload, fully 75 percent of the soil would demand that nutrient. When the calcium is fine as talcum powder, its uptake is rapid and secure. When calcium is solid as a block of granite, its uptake is slow, thus that requirement of centuries for the lush status von Rohmer found.

Henry Turney is a west-Texas cowman and a longtime leader in bringing grass education to the Texas school system. His deduction —*Virgin grass can be virgin only once* — tells us that grass for the pioneers was so rich only because it was virgin.

Chapter 7

The Mineral Diet

Some 25 years ago I had about 500 head of commercial cows. It seems I would have a minimum of 200 calves with *E. coli* scours. I wanted to know why this was happening. I found that *E. coli* scours were there because of a low immune system. On the face of it, this was not a profound observation but tentatively lead in every direction from soil under the pasture to the nutrients supplied in the mineral box.

I cannot claim to be much of a history buff, but I have read about Custer's demise and the awesome toll taken on his horses at Fort Randall. Western stockmen have been afraid of selenium poisoning since 1857 when cavalry horses at Fort Randall, South Dakota, became sick while grazing pastures near the post. Animals die when they ingest too much selenium while foraging. Blind staggers is an advanced form of poisoning, and it often cut into the fitness of the horse supply on the frontier, Custer's horses included, during the dry summer of the Indian Wars. Less than one-fifth of an ounce in a ton of hay prevents nutritional deficiencies in livestock. A deficiency causes degeneration of the liver, pancreas, heart or muscle tissue. But as noted, selenium is also highly toxic. Ranchers often use the term, alkali poisoning.

Certainly soils that produce range vegetation high in selenium are common in some fifteen Western states.

Essential and Toxic

Essential and toxic seems to be a common denominator among the traces. They are essential. The curve then meanders to sufficient and over the hill as toxic. Such nutrients are said to be under homeostatic control, used when needed, excreted when not needed unless the system is overwhelmed.

I select selenium as the opener in this discussion of the mineral box because of an experience over a quarter century ago. *Escherichia coli* became an open sesame for my greater understanding of the Creator's fine tracking system.

A Blood River

An increase in blood acidity or an imbalance of alkalinity will author alarming symptoms. Blood will draw on all reserves of tissue to maintain its slightly alkaline reaction. It can only correct effects with an available supply of alkaline bases, lithium, calcium, sodium, potassium. All must be available in a rather fixed supply to prevent metabolic disturbances. Copper and iron are essential in supporting the blood oxygen supply.

While science finds the number of essential minerals from time to time, it must be noted that the "essential" roster steadily increases. Each element named in this chapter has a function to perform.

Some 50 years ago, credentialed speakers told Friends of the Land why boron, copper, zinc, cobalt, magnesium, molybdenum, iron, etc., were absolutely mandatory in cultivated fields as a prelude to animal and human health.

Jonathan Forman of the *Ohio State Medical Journal,* an associate of William A. Albrecht of the University of Missouri, noted that almost all disease — animal and human — could be erased with well-mineralized farm production.

The spectacle of a Porterhouse steak being practically nutrition-less was cited over 80 years ago, and has worsened since the development of the confined feeding system.

William A. Albrecht and Ira Allison, M.D., studied Bang's disease, or brucellosis, during Albrecht's tenure as Chairman, Department of Soils, University of Missouri. The same disease afflicts human beings by a different name: undulant fever. All forms are a result of deficiencies in trace nutrients. Allison found corresponding deficiencies in cows' milk and in soils used for grazing. By providing the animals with the missing trace elements, he literally fed brucellosis out of the animals. At the time Albrecht said, "This so-called infection, *Brucella abortis,* is about as infectious as the stomach ache." Thus, at a time when officials were putting irradiation on the drawing board, illiterates in veterinary medicine were killing the beast rather than feeding the disease into oblivion. This became the industrial model replacing husbandry. This business of whipping Bang's disease rather than terminating the herd was rejected, the lessons contained in the exercise forgotten by most owners.

Transported to the human realm, the same protocols controlled undulant fever, backache, arthritis, fever, constipation and natural depression, eczema. The milk from cured animals passed along the trace minerals. The elements involved: magnesium, cobalt, copper, zinc.

This discovery made well over three-quarters of a century ago was expected to reduce the use of drugs as a "futile effort," noted G. H. Earp-Thomas, a pioneer in trace mineral studies. In my dairy, clients reasoned the same idea would work for them with their mastitis problems. Drugs, after all, merely dull nerves crying for relief.

In the animal, as in human beings, it is the blood that invites our attention once we have exhausted the signs, symptoms and measurements. Under a microscope, it is possible to watch the circulation of blood in, say, a frog. Under the aegis of health, blood cells behave as a clean up crew, scrubbing and removing wastes. As microbe or toxins are introduced, phagocytes go into battle formation. They take on the posture of a sort of biological clean up crew entering the system, digesting the invader. That same blood smothered by a toxin gives battle to the phagocytes. When the battle goes against the phagocytes, microbes travel the river of blood and will close down neutral antibodies as defense mechanisms.

It takes the trace minerals to neutralize the toxins and reactivate the phagocytes. From our point of view, it is a wonderment to watch those defending soldiers find food for their own survival, then return to the attack, the battleground a microscope slide, the bloodstream a pristine river, or a gutter.

It is up to various organs to make antibodies. They function properly only under conditions of proper nutrition. It is up to pastures and crops to provide that nutritional value.

A Quick Inventory

Trace nutrients stood indicted. A quick inventory revealed that all the minimum requirements were being met except selenium. I reasoned that raised selenium would erase the problem. The day a calf was born I injected 2 cc of MU-SE. After that I did not have one case of *E. coli* scours. Needless to say, each calf had to be caught, not a problem for young men who enjoy sprinting, but age was taking its toll. It came to me that selenium could be added to every 50-pound bag of mineral supplement. This worked. I not only eliminated *E. coli* scours, I seemed also to annihilate foot rot. Retained placenta disappeared as a herd problem. I have not had a mastitis problem. My dairy clients reasoned the same idea would work for them with their mastitis problems, and it did. With selenium, they lowered their incidence of mastitis and their somatic cell count.

Crossbreeding

Crossbreeding is to engage in hybridization, marrying two breeds to create a hybrid. Two breeds cannot be mixed to create a breed unless the cowman has 40 or more years to spare.

Two breeds put a hybrid on the other side of the equal sign. A hybrid has a part of the reproduction efficiency. In crossing two breeds, there is a consequent loss of 20 to 30 percent efficiency.

The Hereford bull and Angus cow accounted for the Black Baldy. Scientific terms tell us that crossbreeding produces *heterosis,* a euphemism for hybrid vigor. Seed stock producers talk a lot about heterosis, a highfalutin' way of saying, "I'm using hybrid vigor."

Hybrid vigor is for the commercial man, not for the seed stock producer. Whenever the seed stock producer is breeding for heterosis, he's pushing the envelope, in fact cheating. Those bulls will not perform. Heterosis creates faster growth, but lower quality meat. The real purpose of crossbreeding is to create a terminal cross.

All the above notwithstanding, there are few absolutes. If you have crossed a Hereford with a Black Angus to produce a Black Baldy, and your purpose is to create a breed, then go back to an Angus bull to produce a three-quarter blooded animal. If you elect to keep the new heifer, go back to a Hereford bull and the progeny will be another three-quarter blood. This represents the greatest benefit out of hybrid breeding.

Real attention to the quality of the bull at each breeding makes production of quality meat a working reality. A shortfall in the process means mediocre commodity meat. The progeny in crossbreeding should exceed both parents.

A Jersey bull on a Holstein cow with another Jersey will create a three-quarter blood, but there is no place left to go when a Holstein is part of the mix. Go to Ayrshire or Brown Swiss and now you have a mongrel — something more than the mule — with little pride of ancestry or hope of excellent progeny. Mongrelization reduces performance.

The Necessary Elements

There are a few Holstein that qualify phenotypically. With such cows and Friesian semen, reversion to abandoned quality could take place quickly. A good line of milk cows would be the result. I know grazers who are mixing Simmental or Gilbey (half bloods) to create a body type.

Single-Factor Analysis

My experience with selenium was a real eye opener. Grass didn't deliver the goods simply because it was there. The soil colloid was supposed to be loaded with 65 percent calcium, 15 percent magnesium, less than five percent potassium and a small percentage of all the trace nutrients. But this was the laboratory speaking, and those printout sheets paid too little homage to those billions — maybe quadrillions — of unpaid microbial workers that did not function in chemicalized, abused or any worn out soils with or without pasture cover.

The calcium-magnesium connection is essential because a proper ratio is required to maintain tissue and bone structure. The libido of a bull relies on the correct balance and usually a well managed soil best governs the correct ratio. There is one event. It takes grass to uptake nutrients in a size suitable to animal usage. This has been calculated to be in the ion or angstrom size, much smaller than the micron size common to minerals supplied by purveyors to the mineral box. In short, forage delivers nutrients that are easily available to cells and protoplasm. Minerals in the box must first be assaulted by rumen bacteria and "made ready" for assimilation.

I cannot overstress this calcium-magnesium connection. A shortage of magnesium causes calcium to migrate into soft tissue, and with that transaction the energy level goes down. A calcium-magnesium imbalance is the aging factor in action.

Calcium to Phosphate

The calcium to phosphate ratio should be about 1.8 to 1. Soils govern ration construction. Soils determine variances and deficiencies, as proprietors of untended pastures sometimes discover and sometimes ignore. A magnesium deficiency reduces the calcium mobilization into the blood.

The key to reading the true status of the cow is blood work. Indeed, the bottom of the problem frequently yields to no other approach. Blood work literally screams out the fact that excess calcium reduces absorption of almost all other elements. Also, pH of water often presents trouble. A high pH can be constructed with excess calcium.

A Directory of Ailments

That dictionary of ailments common to cattle maintenance often seems to have a common denominator. Osteoporosis is a herd given when calcium is interdicted, weakness of legs, subchemical hypocalcemia, retained placentas — all have common causes according to the plane of observation. The common denominator usually breaks down and disintegrates as finer points of reference are revealed.

Thus we have to conclude that sub-causes require calves to be stillborn. The major thing I see is the absence of red blood cells. That is also why dumb calves, slow calves, lethargic calves appear, low red blood cell counts being implicated. Without red blood cells to transport nutrition to various parts of the body, resultant signs and symptoms exhibit themselves.

Two Studies

There are at least two books that ought to have biblical status on the shelves of all cow-calf producers. One is Andre Voisin's *Soil, Grass, and Cancer.* The second is *Eco Farm: An Acres U.S.A. Primer.* Both cover the fine points of soil management, pasture included, and both sketch the connection between individual

Philosophy of biological science

What one does not know

What one knows

"The value of what one knows is doubled if one confesses to not knowing what one does not know. What one knows is then raised beyond the suspicion to which it is exposed when one claims to know what one does not know."

— Schopenhauer

nutrients and herd health, sire prepotency and calving ease. Voisin cited 50 examples from his lectures.

It stands to reason that with a trace mineral key necessary for enzyme function, a shortage of such nutrients or a nutrient imbalance — can alter the health profile. Moreover, industrial and agricultural poisons adversely affect enzyme function, but citing industrial and agricultural poisons does not excuse the mischief caused by hormones, preservatives, heavy metal contamination and inorganic compounds. Description of enzyme function indicates the arrival of nutrition. Many wise old cattlemen reason that multiple inoculations destroy enzyme functions. Poor digestion almost always incubates disease processes. But with the traces in place, the system's enzymes carry out repair and regeneration functions.

Over the years I have assembled a layman's inventory of connections. I now pass them on to those who are willing to read and learn and observe. This much stated, it becomes obvious that any form of preservation of feed inhibits the full function of enzymes.

A deficiency of copper or zinc or molybdenum, can be a consequence of hormones, depleted soil, salt fertilizers or a lack of enzymes. The cause of the curse can be sorted out by signs and symptoms. Toxicity canceling out any enzyme function is a reality and also an elusive connection much like a short in a wiring system. Nutrients delivered via the agency of grass and forage rely on enzyme activity. Indeed, it is this activity that annihilates a deficiency disease, whatever its nomenclature.

In the human food chain, you find fortified minerals. As in animal husbandry, many minerals of a micron size, in the absence of powerful enzymes enabling absorption, actually cause toxic conditions. Cell osmosis cannot occur before particle sizes are achieved that match the size of minerals pulled out of the soil by plant life.

Nutrients that do not furnish food for cell osmosis simply migrate throughout the biological system, often causing mischief known as heavy metal diseases, arthritis included. Too much copper, for instance, can get into the bloodstream, overload the system, and cause enzyme shutdown and digestive failure. In turn, magnesium can form a complex the copper usually available in grass as to cause what Voisin called enzootic ataxia, a kissing cousin to bovine spongiform encephalopathy, or Mad Cow disease.

The pH Factor

Before I recite settled aspects of nutrient imbalance, it seems appropriate to mention pH, meaning parts or potential hydrogen. When the hydrogen level gets too high, acetonemia becomes the observed result. If there is a shortage of hydrogen, the syndrome would be called scurvy in human beings. Ascorbic acid, or vitamin C, is hydrogen in a useable form.

Too much hydrogen replaces oxygen. With oxygen reduced in body fluids, the oxygen content of the entire system is reduced, creating a shortage of cellular oxygen. As pH grows lower, the enzyme function also is lowered, setting off a chain reaction. An alkaline environment is generally a requirement. Herbivores do not have the same requirement of, say, carnivores, a blessing bestowed on them by their appetite for grass.

Copper kills all parasites and intestinal tract worms.

Viral Infections

Viral infections, parasite infestation, and an inventory of nutritional deficiency diseases are a consequence of pH out of balance.

There are nutrients that can be tagged with specifics. Molybdenum latches on to waste hydrogen for removal via urine. And yet the molybdenum in pasture forage is often negligible. Calcium, as mentioned earlier, adjusts pH of foods undergoing digestion. It governs the passage of nutrients into cells and escorts toxins passing out of cells. The late William A. Albrecht wrote whole books on calcium, and one that forever merits rereading is *Soil Fertility and Animal Health.* For an abbreviated bottom line, let it be stated that calcium protects animals from viruses, parasites, molds, fungi and the umbrella for all diseases, cancer. Correct calcium and magnesium supplementation is both the flywheel and the balance wheel of herd health management, and the correct indicator for management of pasture forage.

Zinc stops oxidation damage. Zinc offers certified cancellation of anabolic bacterial and viral health destroyers. This means zinc is selective, always favoring good bacteria, always annihilating bad bacteria. Certain pathogens will not bow to zinc, but zinc debilitates the agent just the same. It transports itself via all the fluids of the system. Of special interest is zinc's role in semen production and fertility.

Sulfur, *et al*

Sulfur is almost always available in green plants. Its absence in a pasture signals a serious pasture deficit. Sulfur delivered by plant life is easily digested with the aid of liver and enzyme functions. I mention this because while sulfur is essential, it cannot be harnessed without vitamin C. But vitamin C relies on the availability of copper. And copper, in turn, needs zinc as an activator. There seems to be a terrible complexity and interdependence, as the following inventory of shortages and excesses readily suggests.

G. H. Earp-Thomas

Of all the scientists who have taken trace nutrients apart and put them back together again, G. H. Earp-Thomas stands above them all. He had developed inoculants for legumes before unraveling the mysteries of the trace minerals. The bovine, not unlike the human being, cannot enjoy good health under conditions of mineral shortage or marked imbalances. Depending on recessive or dominant characteristics, the animal should experience none of the disease anomalies charted below. Organs should not become depleted and diseased tissues should not materialize. Earp-Thomas found that his New Jersey-based cows were deprived of cobalt and other trace minerals by the nature of the soil.

In human medicine, and to a great extent in veterinary medicine, doctors use anodynes and analgesics or narcotics to quiet the results of disease conditions and to numb the pain. But drugs do not supply the missing elements which cause the metabolic mischief.

There is an animal protein factor which is associated with vitamin B-12 production. Vitamin B-12 is constructed in the first stomach of the cud-chewing animal via synthesis of microbes. In the absence of feed, as forage, the essential bacteria fail with a consequent loss of the animal protein factor necessary for creation of vitamin B-12. In extreme cases, the animal dies, and meat protein lacking the "protein factor" transfers anemia to the human consumer.

Vitamin B-14 prevents the growth of the disease cells and hastens the growth of normal cells. This vitamin relies on a trace of cobalt and phosphorous. Both of these vitamins are essential to human health. Earp-Thomas found that only six ounces of cobalt in a New Zealand acre made the difference between health and illness of cattle and sheep grazing those acres. The chances of a suitable cobalt level attending bunk feeding with Bt corn, hybrid silage, and metered feeds for confinement feeding — and an absence of grass — hovers near absolute zero.

We use closed-compartment thinking to remind cattlemen of the superb value of calcium, iodine, magnesium, and iron, but we should not neglect the metaled nutrients that key enzymes. USDA tells of 14 or 16 or 18 essential minerals, the number shrinking and enlarging itself according to the study cited. Don Jansen, the heir apparent to the work of Maynard Murray, the sea solids specialist, holds that the animal and the human being rely on all the minerals on the Mendelyev Chart, an array available only in sea water and in fertility pastures.

The Awesome Roster

The awesome roster of ills that animal flesh is heir to must rate consideration, else it becomes too easy to think of a remedy in terms of crises medicine and not in terms of good soil management. William A. Albrecht used to say, "the quality of bones determines the quality of the horse because bones depend on breeding and the quality of feed. The quality of the feed, in turn, depends on the efficiency with which the plant factory uses air, rainfall and sunshine to bring bone-making minerals along with others — from the soil into the vegetative bulk. We feed our horses and other animals accordingly as we provide fertile soils."

If Albrecht was right, and all the evidence suggests that he was, then the list of animal ailments presented here all call for one remedy. "We may well give emphasis to feed quality for disease prevention, rather than go to the veterinarian for a cure," Albrecht said.

When the *Acres U.S.A.* editor served as assistant publisher of *Veterinary Medicine* magazine, the following notes were kept within reach. *Veterinary Medicine,* after all, dealt with crisis problems, and this meant recognition of signs and symptoms more than knowledge of the first cause. Here are a few problems and what the veterinarian has to say about them.

Acetonemia

This condition is akin to low blood sugar in human beings. Dairymen often call it ketosis. Heavy milker dairy cows are often affected by acetonemia during lactation periods. It can be spotted by watching for an off-flavored milk, a sweetish, offensive breath, bellowing, insane actions, weakness, trembling, sometimes collapse, loss of appetite, or listlessness. Indigestion and constipation also figure in the blanket roster of signs and symptoms associated with acetonemia. The best milkers require extra nutrition and any deficiency is greatly emphasized.

Acidosis

This is a pregnancy disease of sheep. It is also called acute hepatitis and ketonuria. Ewes about to twin are often victims. It is important in sheep growing areas because the condition is associated with a 90 percent fatality factor. Affected sheep can be spotted because they stand behind the flock, grind their teeth, urinate often and tremble when forced to exercise. In the latter stages they refuse to eat and drink and lie down on their breastbones, head to one side. Upon postmortem examination the liver turns up yellow, crumbles easily and has a lot of fat. Kidneys are pale and softened.

Anaplasmosis

It affects cattle, albeit seldom very young ones. Signs and symptoms include anemia, jaundice, frequent urination, rough coat. Death usually takes from 25 to 60 percent of the older animals in a few days. Those that recover continue to carry the disease. At necropsy, skin, teats, and mouthparts exhibit a yellowish color. The liver is enlarged and yellow. The spleen will be soft and not unlike blackberry jam. Gall bladder contents will have a dark green hue and exhibit the consistency of cool gelatin.

Anthrax

This hunts out almost all warm-blooded animals of all ages, man included. Cattle, horses, sheep and goats suddenly die in acute cases. Others run high temperatures, suffer failing appetites, have bloody discharges from natural body openings and exhibit soft swellings that pit upon pressure. Hemorrhages under the skin are common, as are hemorrhages in the throats of swine. Fully 90 percent of all infected animals die. It may take soil processes more than 30 years to rid itself of this disease.

Atrophic Rhinitis

Often a farmer will notice young pigs sneezing and discharge coming from the eyes. These are nasal hemorrhages and irritation of the nose. At necropsy snout and face bones will be observed to be wasted away. No fever here, just a death rate of 20 to 30 percent, enough to rip the profit right out of a livestock operation. This one is a real profit stealer because symptoms are seldom observed before pigs are three weeks of age.

Blackleg

This is a 100 percent killer of cattle, sheep and goats. The symptoms are high fever, loss of appetite, swelling under the skin and shoulders, hip, under breasts, on flanks and thighs. Postmortem findings are more definitive. There will be bloody froth out of the mouth, nostrils and rectum that smells like rancid butter. Bloat comes on shortly after death.

Bluetongue

This one affects sheep of any age, and delivers death between 10 and 40 percent of the time. As the name implies, it starts with inflamed mouth and nose, frothing and labored breathing. The inflamed parts become blue, and there may be ulcers and bloody spots as well as a nasal discharge changing to catarrh with crusts on the upper lip. Necropsy reveals blood and fluids in the lungs as well as muscular degeneration.

Bovine Spongiform Encephalopathy (BSE), or Mad Cow disease

This disease has not surfaced in the U.S. although the human variant, Creutzfeld-Jakob disease, has been noted. The conventional wisdom names the prion as the causative agent and protein bypass from previous generations of animals as a foundation problem, cow casualties as the transport mechanism. More likely, phosphate contamination that inhibits the uptake of copper causes normal prions in the blood to become rogue prions, crossing the blood-brain barrier to create the syndrome. There is no recognized cause.

Brucellosis

Better known as Bang's disease, brucellosis has a second tag, abortion. It affects cattle, hogs, goats and human beings. Signs are abortion, retention of placenta and reduced milk production in cattle. Hog symptoms are abortion, arthritis and inflammation of the testicles.

Coccidiosis

This one is sometimes called red dysentery. It affects cattle, hogs, sheep, goats, and poultry — any age. Very young calves and pigs in particular are affected. Outside of general unthriftiness, there is always bloody diarrhea, anemia and general emaciation. The death rate is not high, but the economic impact is. Coccidiosis exhibits unmistakable gross pathology at postmortem examination: a rectum wall two or three times too thick; contents of large intestines and rectum are chiefly blood.

Cholera

Cholera affects hogs of all ages bringing death nearly 100 percent of the time. In a highly acute form animals just die suddenly, no symptoms. In the acute form, there is a loss of appetite, general depression, fever, purple patches on the abdomen and ears. A discharge often makes eyelids stick together. Because of a weakness in the hindquarters, affected animals walk with a wobbly gait. Examination of organs after death serves up signs, but these can be mistaken for other diseases. Included are a spleen full of blood that is darker than usual, kidneys with pinhead hemorrhages, and raised ulcers in the large intestines.

Diphtheria

There are several handles for this disease condition, namely necrotic stomatitis, gangrenous stomatitis, necrotic laryngitis, malignant stomatitis and sore mouth — all affecting young suckling calves. Several symptoms follow general depression, notably drooling, swelling at the side of the throat, wheezing and coughing. There is almost always a yellowish or greenish-yellow sticky discharge from the nostrils. Odor from the mouth will be quite offensive. Tongues stick out. Death rate is high. At necropsy a cheesy, grayish-yellow mass will be present in the upper windpipe. The same mass is often present in the stomach, lungs, intestines and liver.

Encephalitis

This disease condition is also known as listerellosis and circling disease. It is found in animals with inadequate nutritional support — sheep, goats, cattle, hogs at any age. As the popular name suggests, symptoms include staggering, pushing into fences, circling, and general paralysis. Hogs in particular exhibit hind end dragging, trembling and paralysis. Practically 100 percent of affected animals die.

Equine Encephalomyelitis

This is also called sleeping sickness, and it affects horses, mules and human beings of any age. The death rate is very high.

In all cases fever is followed by sleepiness, grinding of teeth, a wobbly gait and a general difficulty in swallowing and chewing. Sometimes animals become wild and unmanageable. Lips, tongue and cheeks are often paralyzed. Sleeping sickness is often accompanied by pneumonia, gangrene or edema.

Foot-and-Mouth (Hoof and Mouth) Disease

It is caused by the aftosa virus. Some dozen strains have been identified. Aftosa appears to be an opportunistic organism. It can infect an entire herd. Some strains are mild, some virulent. Poor and improper nutrition has been implicated. Hydrogen peroxide annihilates this virus, especially when used as a preventive measure. There have been no cases in the U.S. since the late 1920s.

Johne's Disease

Also known as paratuberculosis and chronic specific enteritis of cattle, horses, sheep, goats and deer. Johne's disease starts with a general loss of condition — thirst, diarrhea, rough coat, dry skin, but no fever. In the final stages animals simply refuse to eat. Signs and symptoms are difficult to discern clinically. Much of the same is true of necropsy findings in younger calves, which exhibit nothing more than swelling of the small intestines, often in a small area. In older animals the ileum, cecum, large intestines and rectum are thick and have red patches. The thickened areas can be up to five times the normal thickness.

Leptospirosis

It affects cattle, hogs and horses, killing at least 35 percent of affected animals when they are bovine. In dairy animals there is a drop in milk production, milk becoming thick, yellow and blood tinged. During pregnancy the disease causes abortion. In hogs the usual symptoms include circling and meningitis. Calves exhibit fever, prostration, labored breathing, anemia, jaundice, and red urine. As one might expect, necropsy findings always include urine in the bladder approximately the color of port wine. Anemia and jaundice are also typical. Kidneys have white spots or reddish-brown ones.

Malignant Edema

It affects horses, hogs, sheep, cattle — any age. It is also known as gas phlegmon and as a braxy in sheep. It is 100 percent fatal. Veterinarians usually associate it with trauma such as nails, castration, docking, or shearing. High fever, loss of appetite and swelling touch off the symptoms parade. Pressure causes a thin, reddish fluid to flow from swellings that, in any case, make a crackling sound when touched. Swellings are usually found in the lungs and elsewhere, as in blackleg.

Milk Fever (Parturient Paresis)

This disease condition is usually associated with dairy animals, although female sheep, goats and hogs can be affected. It occurs after newborns are delivered. Symptoms include refusing to eat, trembling, staggering. Without treatment such animals die within hours.

Navel Ill

This is a problem with newborn horses, sheep, cattle and pigs. It is sometimes called joint ill because hot swollen joints are a common symptom. The death loss is usually near 100 percent when the condition comes on within a few hours after delivery. Otherwise the death rate drops to between 30 and 75 percent. Necropsy findings include abscesses in the spleen, liver or lungs. The navel is generally inflamed. Joints often contain pus, particularly hock and stifle joints.

Necrotic Hepatitis

This is called black disease of sheep. It has no symptoms except that death comes swiftly. Sometimes a bloody foam will run from the nose. Bones of the snout and face are often wasted away.

Pink Eye

Sometimes known as infectious keratitis and infectious conjunctivitis, this one affects cattle and sheep of any age. There may be occasional blindness and some few deaths, but the main result

Copper kills all parasites. intestinal worms
Selenium - E coli

is loss of milk production or unthriftiness. In all cases eyelids become red and swollen. A yellow deposit forms over the eye.

Pneumonia

Inflammation of the lungs affects all warm-blooded animals of any age, human beings included. It is often associated with other disease conditions. In animals as in human beings, symptoms include dullness, high temperature, rapid breathing, and hard pulse. There is often a discharge from the nostrils. Wheezing or gurgling sounds can often be detected in the breast. Cattle breathe through the mouth and extend their tongues. When examined after death, lungs exhibit reddish or grayish-red patches. Air tubes of the lungs are filled with yellow or gray pus.

Red Water Disease

This affects cattle and sheep of any age. Bowel movements are scanty, then bloody diarrhea follows. The most common symptom is foamy urine the color of dark red wine. Death comes within 24 to 36 hours fully 95 percent of the time. Severe anemia and jaundice are part of the clinical picture at necropsy.

Scrapie

Scrapie of sheep is fatal nearly 100 percent of the time. Postmortem examinations yield nothing unless fine pathology instruments are available. In the field a farmer might note fine tremors that produce a nodding movement. Intense itching starts on the rump, then travels over the entire body. Emaciation follows. Sheep tend to step high in trotting. This condition runs its course in six weeks to six months.

Shipping Fever (Hemorrhagic Septicemia)

A lot of farmers call shipping fever stockyard pneumonia or slobber disease. The death rate can reach 10 percent, sometimes more. Cattle, sheep, and hogs are affected. In almost all cases there is a hacking cough, watery eyes, a discharge from the nose and that swollen tongue that causes animals to slobber and drool. On postmortem examination lungs turn up phlegm-coated. Lungs have a reddish serum. Lymph and throat are swollen.

Sore mouth

This disease condition of sheep and goats affects chiefly young lambs and kids. As a precursor of other disease conditions, it can account for very high economic losses, often up to 50 percent. Symptoms are reddening and swelling lips and gums, and blistering of these parts and the swollen tongue. Pus breaks out in a few days. Scabs from on the resultant raw spots. Scabs usually fall off in a few weeks.

Strangles

This is really distemper, and affects horses and mules. The death rate is usually low, about five percent, and is restricted to colts over six months of age, and horses between two and five years of age. A reduced appetite, a discharge from the nostrils, snorting and coughing, plus swelling under the jaw and in the throat, all mark the etiology of this condition. The swollen area becomes filled with thick yellow pus, and this interferes with breathing — hence strangles.

Swine Erysipelas

It affects hogs and human beings, primarily, but is often a problem in sheep and turkeys as well. The death losses are very high in acute cases. In a way, swine erysipelas resembles hog cholera. Animals in seemingly good health die quickly. Some animals have reduced appetite, a stilted gait, red patches on the belly, and suffer vomiting, constipation, then diarrhea. Postmortem examinations generally reveal swollen lymph glands and spleen, hemorrhages in the stomach, intestines and kidneys. Joints often contain fluid. Joint bones are frequently wasted away.

Swine Influenza

This one often infects an entire herd. It starts with loss of appetite, coughing, discharge from the nostrils, and progresses to red eye, labored breathing and high fever. Hogs squeal mightily when handled. Postmortem examinations always reveal hemorrhages in several organs of the body. Lymph glands contain blood. Body cavities and lungs contain fluids. Bleeding around and in fatty tissue is symptomatic.

Tetanus

This is simply old-fashioned lockjaw. All warm-blooded animals, *Homo sapiens* included, can be victims. The problem can almost always be traced to cuts or puncture wounds. Chewing and swallowing become difficult. Muscles attain an alarming rigidity. Animals take a stiff-leg stance, tail raised. The mortality rate is extremely high, sometimes 90 percent in the case of Eastern-type infections.

Trichomoniasis

This condition of mature cattle results in abortion. It is different from Bang's disease in that the dead fetus remains in the uterus for some time. Sexually mature animals generally survive, but the condition takes a toll in fetuses.

Typhus

Infectious necrotic enteritis or "necro" is a disease condition that affects two- to four-month-old hogs and delivers a high death rate. At the onset there is fever, loss of appetite, diarrhea and general unthriftiness. It is easy to mistake pig typhus for hog cholera. This is important because vaccination with the wrong serum will not halt the death march.

Vesicular Exanthema

This is an economic disease, one that costs in heavy losses of young pigs. It can be recognized by blisters on the snout, in the mouth, between toes, the soles and dew claws, which are filled with a clear fluid. Unless ruptured, blisters become infected but usually heal in a few weeks.

White Scours

Sometimes known as infections diarrhea or acute dysentery, white scours affects calves up to five days old. Fully 90 to 100 percent of badly infected animals die. The chief symptom is a yellowish to white foul-smelling diarrhea. Often temperature drops below normal. Examinations after death reveal a reddish serum in the body cavities. The liver, spleen and kidneys are almost always at least partly wasted away. The digestive track is always inflamed.

X Disease

This disease of cattle is also known as bovine hyperkeratosis and it affects all ages, delivering a death rate of near 80 percent in calves that are very young. The death rate usually goes down for older animals. In X disease, raised rounded bumps appear in and around the mouth. There is a watery discharge from the nose and eyes. There is loss of appetite and diarrhea. The gall bladder, liver, kidneys and pancreas are affected. Abortion and mastitis in cows are commonly part of the X disease condition.

Health Out of Control

Even the cursory examination of these disease anomalies suggests that one would be thrice a fool to attempt raising animals. The veterinarians have vaccines and serums and needles to cope with many of these problems, all of them a part of the practice known as crisis medicine. Crisis, however, means what it says — health out of control. Each of the conditions previously noted represents a health profile out of control, or physiological bankruptcy. It may be that some of the problems can't be prevented even with good nutritional support. It is safe to say that 95 percent of crisis medicine can be sidestepped by dealing with the soils first, feedstuffs second, finally with the animals themselves.

Albrecht said it all when he wrote in *Good Horses Require Good Soils.* "When a plant can make much forage yet deliver no seed, the wide fluctuation in chemical composition of the vegetation should become evident. Grass crops that are measured in terms of tons of forage in place of seed yield per acre may be growing on soils too poor to make seed, yet we accept their forage without suspecting defective composition and poor feeding quality. Such soils have mainly a site value and serve largely as plant anchorage. It is this soil's property that makes forages deceptive as feeds."

Albrecht went on to note that "Wayne Dinsmore (an associate) reported that *periodic ophthalmia, commonly called moon blindness, frequently is seen in states east of the Missouri River and occasionally west thereof.* In that statement he revealed the possibility of the dominance of deficiency diseases within regions of heavy rainfall, or on the humid soils and their decrease with less rain, or regions of arid and less leached soils." Or, as an octogenarian Hereford breeder in Missouri once said, "They aren't doing on this land what they did here 50 years ago." In short, as nutrients run out or become complexed, the general imbalance conspires to debilitate the animal. Pathogenic organisms are opportunists. Much like weeds or insects, they select the weakened body for their handiwork.

Viewing this awesome list of disease problems — all of which can cut the profit picture to pieces — one must ask the obvious question: *Just what does it take to keep an animal in good health?* We have already suggested that crisis medicine is a poor answer. Then what is a good answer?

Warren Spring of Milledgeville, Illinois, once provided and codified an uncanny summary of research work that consumes countless pages of scientific literature. "What you should have for an animal is a balanced ration," he said. "This is usually taken to mean so much protein. Yet the first thing you have to get in your mind when you balance a ration is that the most damage you can do will be with an excess of anything."

This is true. When an animal has excess, it has to borrow from the energy source to get rid of it. Ruminants, for instance, live on the bacterial crop in their unique chambers. The more efficient the production and the better the crop, the more near optimum is the nutrition of the ruminant. What is optimum? Optimum is not a minimum daily requirement, or being certain that there is plenty of a certain element, but the best possible level of each nutrient in respect to other nutrients required by rumen micro flora.

A well-grown plant, even though stressed by the hazards of nature, will supply the best possible level of nutrients — all biologically processed into the colloidal content of stems, leaves and seeds. It is this form of nutrient material that will supply the entire animal system with the digestible ingredients that assure ingestion into the colloidal fluids and blood supply. At least 70 percent of the mineral intake must be in colloidal form for optimum health and body function. Too many of our synthetic nutrients never reach the blood system, and more often than not they merely pass through the animal.

If we are to avoid health problems, we must grow or select food materials that are physiologically mature and ripe, and these materials must be obtained from plants that are grown on soils in basic equilibrium. It is such feedstuffs that stimulate optimum hormone processes essential to support of the biotic system involved. This balance, we have pointed out, is best achieved with an array of essential amino acid nutrients and protein of the right character.

A Mathematical Formula

Some parts of nature's equation have been reduced to a mathematical formula by Philip C. Anderson and Janet L.C. Rapp. Their breakdown was first proved out at Oklahoma Panhandle State University, the Wisconsin Alumni Research Foundation and the University of Nebraska.

Any system for feeding farm animals properly has to deal with quality. It has to scale the measurable intervals between severe deficiency, hidden deficiency, optimum excess and severe excess. In terms of a pie chart, here are the basics — just five simple isolated groups of elements extracted from each other. Unless they are brought together by the systems of life, energized by sunlight, processed by photosynthesis in the leaf of a green plant, assembled into molecules of nutrients by "cell workers," and packed into a carbon-oxygen bundle by the forces of hormone and enzymes in a living plant, they cannot sustain higher forms

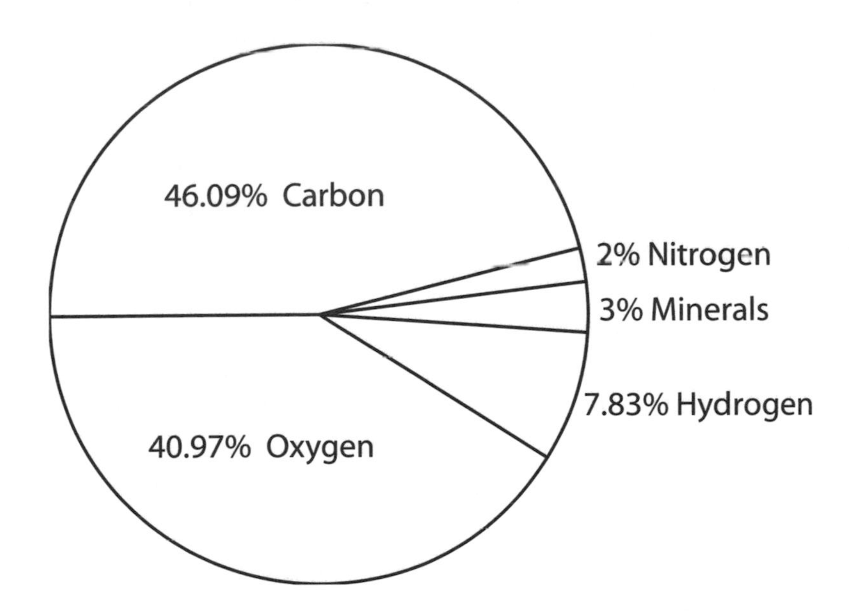

of life. And if any of these systems are short-changed either in form or amounts of each ingredient, consequences of malnutrition will be the result. Can we simply say that animal health depends upon soil that is healthy, and a healthy soil depends on a healthy microbial environment? There is a biotic relationship that is essential to life, and man — and the production of animals — cannot endure a folly of substitutes.

A pet antipathy of many eco-farmers is the feed trade's preoccupation with protein. In 1928 *Morrison's Feeds and Feeding* started reporting protein in feedstuffs. They took samples from all over the United States, with Eastern states weighted heavily. Each year more samples were added, and yet the old ones were not dropped. Thus there exists in *Morrison's* a cumulative figure bearing only a marginal relationship to the fact of crop nutrient content.

For instance, the fiction of total digestive nutrients (TDN) has survived. Yet years ago Purdue found TDN passé. They proved that three corns could test the same TDN-wise and yet litter mates would do differently on it. Second, TDN takes no account of oxygen at all. Excess oxygen in the feedstuffs on a dry matter basis lowers energy. Energy is the flywheel and balance wheel, but not the spark plug, of animal health.

Here are a few notes on feedstuffs. Observe how feed imbalances quickly relate to many of the disease conditions discussed earlier. In a few cases we know exactly which nutrients bring on disease, for instance, brucellosis and grass tetany, a condition generally brought on by too little magnesium in the diet. In others we can only hint at the mix of imbalances that cause pus to flow, urine to turn wine red, scours to debilitate, internal organs to waste away. Moreover, arrival of the synthetic kingdom — the herbicides, the pesticides, all the toxic genetic chemicals — ape and mask disease conditions much as syphilis imitates many of the common disease conditions in man. Unfortunately modern science often pauses when it discovers a proximate cause, and forgets to pursue the foundation cause — an oversight that has kept the white light of analysis away from nutrition far too long.

Content of Ration

Dry matter and the ash content of a ration govern the solubility value of ration and the rate at which it moves through the digestive tract. Dry matter should not be less than 20 percent for the ruminant. The ash content, which measures the oxides of minerals in the ration and provides an index of fiber content, should be between four and six percent.

As illustrated above, the carbon content should be 46.09 percent, according to Anderson and Rapp. Carbon governs the need for hydrogen, oxygen and nitrogen, excess increasing the requirement for these elements, a deficiency decreasing that same requirement.

Note the posted figure for hydrogen, 7.83 percent. An excess over this figure means the energy value of the ration is in excess. Bloat is a likely result. Deficiency means mucosa disease, watery eyes, a hacking cough, undigested feed in droppings, unthriftiness. It takes little imagination to match a nutritional lack to the specifics veterinarians try to remedy with crisis medicine.

If the oxygen content of a ration is in excess of 40.97 percent to 41.5 percent, animals founder and exhibit a stiffness of gait. The depth of the body becomes shallow or "tucked up." Roughage and grain in droppings increase. Milk and butterfat production drop, and the rate of gain in beef animals drops. If there is an oxygen deficiency, butterfat production may stay up, but milk production drops. The rate of gain is slowed. A warning sign can be seen in droppings that stack up like canned biscuits because the ration is not soluble.

The Nitrogen Level

The nitrogen level has to hone closely to that two percent figure, never to exceed 2.88 percent. An excess means ketosis, scouring, mastitis, a general decline in milk production or rate of gain. At the other end of the scale, deficiency accounts for retarded growth, irregular heat periods, poor rate of gain in beef animals and poor milk conversion in dairy cows.

The lexicon of animal husbandry holds to the equation 6.25 x total nitrogen = protein equivalent. Proteins are a major component of every living cell. Protein as a 12.5 percent component of the ration is optimum for all ruminants. But these proteins must be ripe and mature proteins, not just a laboratory measurement of nitrogen content. Too much protein is a contributing factor in ketosis, scouring, poor feed conversion. A deficiency soon translates to retarded growth, poor wool or milk production, and irregular heat periods. Needless to say, problems do not generally arrive as a consequence of single-factor imbalance. Often several parts of the equation are out of whack when animals become the focal point of exotic veterinary talk. Sulfur, potassium, calcium, magnesium, phosphorus, chlorine, silicon and the trace minerals, all figure when they account for an imbalance — either too much or too little.

More on Sulfur

The optimum amount of sulfur should be .14 percent of the total dry matter. An excess accounts for an acid rumen and increases the need for copper. Too little sulfur causes sheep to shed their wool and results in poor development of keratinous tissue such as hoofs, horns, and hair. Excess saliva and watery eyes become a legacy. It may not be possible to tell just which other imbalance joins a sulfur imbalance to trigger anything from acetonemia to bovine hyperkeratosis, but the foundation cause can easily be seen.

Cattlemen's Clues

My co-author, Charles Walters, has assembled some of the clues that link physical anomalies that attend livestock production, the focal point being trace nutrients.

Potassium

East Texas farmers still talk about the potassium shortage in native grasses that once had cattle either dying or becoming shy breeders. In fact potassium should account for .93 percent of the ration on a dry matter basis. Too much slows down bacterial growth in the rumen. Too little results in retarded sexual maturity and difficult breeding. Because carbohydrate utilization is slowed by too little potassium, growth is also slowed.

Sodium

Sodium also has its index. The optimum amount for a ration should be .27 percent of the dry matter. Possibly 90 percent of all rations, even many "scientifically balanced" rations, are low on sodium. Too much, of course, figures in swelling due to water retention and inhibited bacterial growth in the rumen. Too little short circuits utilization of protein and energy. Rough hair coat, retarded growth, poor appetite and poor reproduction all point to sodium deficiency.

Calcium

Calcium is the prince of nutrients in the soil. It should occupy .48 percent of the total dry matter. It happens also to be basic in the animal ration. Too much increases the need for phosphorus, vitamin D-2 and zinc. It decreases the availability of protein, phosphorus, iodine, iron, manganese and zinc. It figures in birth paralysis and depresses the rate and economy of gain. A deficiency, on the other hand, impairs bone growth. It increases the need for vitamin D-2. In the feedlot or on the pasture, the calcium-deficient animal will appear listless, often exhibiting an arched back and a depraved appetite.

Magnesium

Magnesium should be .29 percent on a dry matter basis. In excess it increases the need for phosphorus. As a deficiency, it accounts for grass tetany, a condition that sees animals blinded, turning in circles until balance is lost, and frothing at the mouth.

Phosphorus

Phosphorus walks hand-in-hand with calcium. When calcium is excessive, cattle will eat phosphorus in excess and then excrete both calcium and phosphorus down to the optimum level. Phosphorus, of course, is acid in nature. An excess of this nutrient increases the need for iron, aluminum, calcium, and magnesium. The observed result is poor skeletal formation. Deficiency means poor fat assimilation, either a delayed or missing heat period, prolonged intervals between calving. On the economic front, there is poor rate of gain, faltering milk production, general unthriftiness. Just as an excess of phosphorus increases the need for iron, aluminum, calcium and magnesium, a deficiency of phosphorus can be created by excess iron, aluminum, calcium and magnesium.

Chlorine

Chlorine is another element that must figure in the ration — this one to the tune of .42 percent of the dry matter. As with sodium, excess brings on swelling due to water retention. It also increases the iodine requirement. A deficiency results in loss of appetite and poor weight gain, poor hair coat, hyper-alkalinity. Tetany and death are often a result.

Silicon

Silicon has its optimum figure in the ration, .33 percent. Any excess slows passage of food through rumen. It also decreases digestibility and palatability. A deficiency slows growth and multiplication of rumen bacteria, causes poor fill, and a depraved appetite.

Trace Minerals

In addition to the above, there are the trace minerals, most of which have not been investigated in terms of animal health. Individually and as combinations, they probably contain more answers to health problems than all the medicines ever synthesized by man.

Trace minerals are reported as parts per million (ppm) on a dry matter basis. Optimum amounts of trace elements are expressed as follows:

Iron	100 ppm
Aluminum	60 ppm
Manganese	60 ppm
Zinc	60 ppm
Boron	10 ppm
Copper	10 ppm
Molybdenum	1.0 ppm
Iodine	0.5 ppm
Cobalt	0.5 ppm
Nickel	0.5 ppm

Iron

An excess of iron interferes with phosphorus absorption. Sodium or potassium bicarbonate is required to precipitate iron excess. A deficiency of iron results in anemia. This is most likely to occur in calves because milk is low and little iron passes across fetal membranes. Cow and calf operations can exhibit anemia and are more susceptible to disease conditions.

Aluminum

An excess of aluminum increases the need for phosphorus.

Manganese

An excess of manganese increases the need for iron. A deficiency of manganese, on the other hand, results in leg deformities with over-knuckling in calves, eggs not formed correctly, degeneration of testicles, offspring born dead, delayed heat periods. Shortage is created by an excess of calcium and phosphorus.

Zinc

An excess of zinc means decreased copper availability and interference with utilization of copper and iron, bringing about anemia. A zinc excess also shows up as bald patches and skin disorders (rough skin), a deficiency is created by excess of calcium.

Zinc is absolutely necessary for production of sperm. It also increases the need for vitamin A.

Boron

An excess of boron means diarrhea, an increased flow of urine, and visual disturbances. A deficiency of boron reduces rate of growth as rumen bacteria.

Copper

An excess of copper results in degeneration of the liver. It causes blood in urine and poor utilization of nitrogen. A deficiency of copper is created by excess of molybdenum and cobalt produces anemia due to poor iron utilization. It depresses growth. Other symptoms include depigmentation of hair and abnormal hair growth, impaired reproductive performance and heart failure, scouring, fragile bone, retained placenta and difficulty in calving, muscular incoordination in young lambs, and stringy wool.

Molybdenum

An excess of molybdenum makes copper unavailable. It brings on depigmentation of hair, and also severe scouring. A deficiency of molybdenum is created by excess of sulfur. It slows down cellulose digestion. It accounts for calcium deposits in the kidneys. Chronic poisoning is also an observed result, depending on the level of copper. A deficiency of molybdenum slows down the conversion nitrogen of protein.

Iodine

An excess of iodine means secretion of mucus from the lungs and bronchial tubes. A rapid pulse and nervous tremors also accompany iodine excess. A deficiency of iodine brings on unmistakable signs: young are born dead, or die soon after birth; abortion at any stage, or reabsorption of the fetus; retention of fetal membrane; irregular or suppressed heat period, infertility and sterility; decline in sex drive and deterioration in semen.

Cobalt

An excess of cobalt reduces the availability of copper, aluminum, iron, manganese, molybdenum and iodine — if the excess is severe. Also, the ability of bacteria to convert nitrogen to protein is reduced. Cobalt deficiency means that rumen bacteria fail to manufacture enough vitamin B-12; starved appearance with pale skin; decreased fertility, milk, or wool production. Cobalt is necessary for utilization of propionic acid. Without it, cellulose digestion is sharply reduced.

Nickel

An excess of nickel makes the ration unpalatable. Excess nickel can be reduced by chelated iron.

Vitamin A

This is the anti-infection vitamin. An excess of vitamin A is stored in the liver and in fat tissues. It works against vitamin D. On the other hand, a deficiency of vitamin A results in the following inventory of symptoms: nasal discharge, coughing, scouring and watering eyes due to crying and hardening of the mucous membranes, which line the lungs, throat, eyes, and intestines. Calves' horns are weak. Severe diarrhea in young calves is observed. Redness and swelling occurs around dew claws. There is stiffness in hock and knee joints and swelling in the brisket, increased incidence of mastitis and other udder problems due to drying and hardening of the mucous membranes of the udder. There is also decline in sexual activity and sperm decrease in number of mobility. Loss of appetite is another signal. Zinc deficiency, nitrates, and low ash rations increase the need for vitamin A.

Vitamin D

An excess of vitamin D will result in a deposit of calcium in the heart and kidneys and works against vitamin A. A deficiency of vitamin D has the following symptoms: joints and hocks swell and stiffen, back arches, increased need for calcium and phosphorus, stiffness of gait and dragging hind feet, and rickets.

Vitamin E

A deficiency of vitamin E usually occurs in young animals. Effects include muscular dystrophy (white muscle disease) and heart failure, paralysis varying in severity from light lameness to complete inability to stand.

Vitamin K

A deficiency of vitamin K results in failure of blood to clot.

There isn't a disease condition in animal husbandry that isn't started or sustained by poor nutrition. Likewise, there need be no animal disease if the animals are fed plants that give good nutritional support.

The late W. P. Scott of Naremco, Inc., Springfield, Missouri, said it when he noted that "virtually all disease causing organisms are everywhere in nature, and in all forms of life. Yet we never see infections caused by bacteria or fungi present in all people, all animals, or all birds. Why do some escape? More important, what are the conditions that make a bird or an animal unable to guard its own health?"

John Whittaker was a veterinarian who followed viruses, parasites, protozoa, bacteria and fungi cause-effect relationships his entire professional life. During the last two decades he wrote often and well about mold toxins, those unfriendly fungi that seem to get their go-ahead when friendly fungi in the soil are overwhelmed. Poor soil management helps underwrite this mycotoxin proliferation, *Aspergillus flavus,* for instance, but not half as much as increased use of toxic genetic chemistry.

There are consultants afloat in agriculture who attempt to put all these nutritional requirements in a computer and spin out answers, often with remarkable results. Basic foundation ingredients and the trace nutrients all figure when computer lights flash. Much of the fine-tuning relies on chelation, a basic phenomenon in plant and animal life. When feedstuffs fail to deliver the full balance that health requires, which is frequently the case nowadays, then the next best thing is to turn to the chelated additives.

All this is not to suggest that sound principles in crop production won't produce feedstuff support for animal health. Sound principles will. Gene Poirot, of Golden City, Missouri, kept a cowherd free from disease for decades. Environmental farmers everywhere are doing the same thing without computers or even in-depth knowledge of nature's chemistry. They know that they must reach back to the soil for nature's balance, rhythm and harmony, or the sad notes on which this lesson started become an assured legacy. And they act accordingly.

A Personal Note

I use Albion beef and diary minerals. This is a pasture mineral. MAC-metal amino acid chelates — copper, zinc and manganese designates my favorite.

What is the tag line to health and disease and the mineral box? Dr. Albrecht, quoted above, put it bluntly. An animal walking to the mineral box, he said in effect, is exhibiting desperation. Something is missing in the forage. Better find out what it is.

Chapter 8

Cowboy Arithmetic

If not fixing the problem is a tribal habit among cattlemen, then failure to plug in simple cowboy arithmetic often furnishes the *coup de grace.* The problem often could be fixed by going to the soil. To this end I recommend *An Acres U.S.A. Primer,* now in its fifth edition and available through Acres U.S.A. under the title *Eco-Farm: An Acres U.S.A. Primer.* Another part of the problem could be fixed by dealing with the phenotype of the cow. Unfortunately, to do either of the above might bring on social rejection at the coffee shop. Worse horrors, neither of these steps is likely to win a blue ribbon at the county fair.

Numbers

Our culture seems to close down any realization that a change in the soil system or the phenotype of the animal could enable rejection of the internal parasite salesman, even the mineral salesman.

In 1987, a friend in Kansas concluded that the direction he was going would lead straight to the poor house. He gathered up several head of cattle and had a sale. Then he scoured the countryside. He found 25 cows that fit the profile outlined in the several chapters of this book. He started flushing those 25 cows. He put those eggs in a bunch of commercial cows. By 1995 he had another sale. He sold $1.3 million odd worth of cows out of these 25. That's real cowboy arithmetic. He's about ready to do it again as I set down these lines — using the same 25 cows. But remember, it took a lot of highway miles, travel costs and three years to find these 25 cows.

There is a lesson tucked into this vignette. "I ran mineral salesmen off," he told me. And I can testify that there isn't a mineral box on that farm.

This is possible because those cows are made right and they are functioning for him. Moreover he is located in the high plains, which means calcium-rich soil, an optimum amount of rainfall and a natural mineral mix in the soil, a gift of the glaciers at the end of the last ice age.

I have seen all these cows he flushes. One and all, they have that big gland that comes up the side. He's getting 15 to 25 eggs out of each cow. There is a story about the man who gave up golf because it was so much fun he couldn't make a living. My friend has fun and work harvesting those eggs because the fun makes his brand of cowboy arithmetic serve him.

The environment and the animals get along on native pasture, "the forgiveness of nature, her constant benediction," another Kansan dubbed it. He does not over-graze unless there is snow cover; he finds it unnecessary even to feed hay.

My friend is a standard. He is also independent. He may not need membership in a breeder's association, but his fellow cowmen need him.

The Industry

The industry today has suggested to itself that cowboy arithmetic starts with cows breeding, the linchpin of hybridization. Mixing breeds creates a hybrid, not necessarily a bovine mule or a hinny, but a mongrel. You don't mix breeds to create a breed, at least not in terms of the working life span most of us have. The hybrid loses a part of its reproduction efficiency, about 20 to 30 percent of that efficiency, in fact. Crossbreeding produces heterosis, a fancy term for so-called hybrid vigor.

These are equations that deliver in terms of commercial aims, but this is not a working reality for the seed stock producer. The seed stock producer who breeds for heterosis or hybrid vigor has found a euphemism for cheating his clients. Fast growth and low quality meat is not the ticket for a good bottom line.

The product of cross-breeding should be the terminal cross. This is not to say that, say, a Black Baldy out of a Hereford breed and an Angus cow can't qualify, but I suggest an Angus bull to create a three-quarter blood, this procedure to be followed with a Hereford bull to equal a three-quarter blood. This describes the greatest benefit out of the hybrid vigor paradigm. This system of breeding sequence can produce high quality meat. Anything less results in commodity meat. Progeny should always exceed both parents.

From a cowboy arithmetic point of view, a Jersey bull on a Holstein will produce a good milk cow. A return on such progeny with a Holstein bull, in my opinion, is not possible. I would use a Jersey bull on a Holstein cow, then come back with another Jersey and create a three-quarter blood. Unfortunately, there is no place to go after that. Some dairymen go to Brown Swiss or Ayrshire, but the down-track result is the mongrel.

This Rule

This rule has been covered elsewhere in this book and possibly constitutes a digression since the arithmetic of maintenance and bottom line calculations assumes foregoing givens. Suffice it to ask, does anyone breed and sell cur dogs? I know all absolutes are false. Nevertheless, attention too often fails to answer that question because it has never been asked.

The signal word has been and remains low maintenance.

The pastures are full of high maintenance animals. The average calving percentage in America is 70 calves per 100 cows, this for beef cows. In some states, such as Arkansas and parts of Missouri, 60 percent is maximum. These percentages are deeply embodied in cow-calf production. I have a friend who has created a term, *adjudice,* meaning "after judge," or the opposite of prejudice. I suppose I have some prejudices, but I think most of my opinions are based on delayed judgment.

There are crosses that work well, based on selection and ruthless culling. British White bulls work well while producing some great colors and a tender end product. Angus and Hereford suggest equally excellent crosses for the rail. I cannot endorse any Brahma crosses with any breed. The result is always about ten inches of bone, no maternal traits in the females, not an end-product oriented procedure. The loss of meat is astonishing.

Other Crosses

I am not partial to Limousins. The animal does not seem to possess even minimum intelligence. Some of the Charolais crosses do not work well. The commodity meat producer most likely

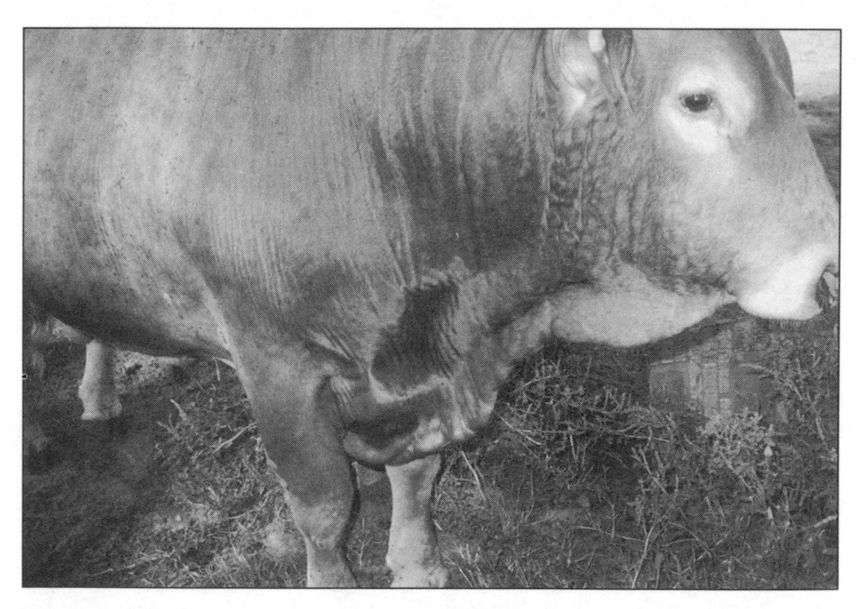

Identifying the thymus via a swirl of hair on the jugular.

does not care, but the producer of excellence on the hunt for premium markets has to do bookkeeping from every angle.

The Black Baldy may be one of the most docile crossbred animals around. Generally, the cross will not have the intelligence of either parent, a definite negative. A half-blood will have only half the genetic pre-potency of either parent. Half translates into high maintenance.

The arithmetic of management argues that everything relates to everything else. Earlier, I pointed out a sign related to the thymus gland. I have a picture, reproduced here, that illustrates the point and invites attention to the management problem of overfeeding.

The thymus gland presents itself as a swirl of hair that looks like a shadow. The dark area near the jugular goes up the side of the head in an extremely healthy cow. It really shows up when the sunlight is just right.

The Pancreas

The pancreas sign along the side tells the cattleman that this animal is four months pregnant or more.

Here is a good horrible example of an animal that is overfed. Note the fat brisket. This cow probably weighs 1,500 pounds. She goes about 16 months between calves. Calves are weaned at about 205 days. When fat builds up around the ovaries and the reproductive track, pregnancy is difficult to achieve.

Over-fat usually means underproduction. Observation and calculations tell me that we grow our calves too fast, that we rape our pastures by feeding large numbers and gains to create a bunch of meat that is not palatable. Consumers are drifting away from red meat protein because too often a T-bone or Porterhouse steak is a bad eating experience.

Too much green plant also impacts on the arithmetic of cow-calf production. Some of that hay production might better be farmed out to a neighbor rather than owning the equipment.

Fat cow brisket.

I've heard the comment, "I'm doing pretty good," and there are words like excellent, profitable, and not too bad. For my part, I love numbers. Accordingly, I've put together numbers for 400 acres, 100 cows, assuming the average soil in the South, and I hope it will enable the cowman to draw the appropriate conclusion. This model will change according to the area, but it should guide the logic and thinking that backgrounds a profitable bottom line.

The Calf

The average calf has seven owners. It travels 1,400 miles from the time it is born until it makes it to a dinner table. There are too or three beef organizations formed recently that hope to achieve *select* as their target norm. The American Hereford Association recently entertained the billingsgate that Hereford select was better than Angus choice, this according to Colorado State research.

Prices change on a pump and dump basis, the blue sheet rigging the value. For my purpose, I'll assign a value of $2,100 to

$2,400 via the feedlot destination, the average carcass weight is somewhere between 750 and 850 pounds.

The price received for 70 calves out of 100 cows (550 pounds each) should net $34,650. Seventy calves at $1.00 a pound — $38,000. Ninety calves at $1.50 a pound would enable a farmer to compete with the rest of the industry. The bottom line would be about $74,000 for that year.

The average carcass through the infrastructure actually delivers about $3.00 a pound by the time it touches base at every stopping point along the infrastructure way. The average calf — plus or minus — is sold at 24 months of age. The average price for a 550 pound calf at 90 cents would be $495. The average cost of keeping a cow for a year, I compute to be $250. (Sad to say, some operators spend $400 a year per calf). Taking those same 400 acres, let's sell half those cows. Let's rotate our grazing Voisin style and do the finishing on grass. That bypasses the infrastructure to you and the housewife. Here the average calf receives no hormones, antibiotics, or chemicals. Consequently, the average calf is never sick. Quit fiddling with those animals with all that junk, keep them on good grass, and natural good health will assert itself.

The average calf travels plus or minus 25 miles to the processing facility under this simplified system. Thus the stress is minimal. The average calves that I marketed were graded select and I sold them all at 14 months of age. I did not finish them or want to finish them. When I started getting 20/100 inches of body fat, my optimum was achieved. My customers were extremely pleased.

The average calf going through a routine grazing sequence would be 18 to 20 months of age before being sold. The average weight of a grass-fed steer should be somewhere between 800 and 1,100 pounds, variations assigned by breed. The average carcass weight should be somewhere between 500 and 700 pounds. But the average price received through direct marketing — assuming the value the infrastructure gets — will be $1,500, all according to this model. Remember, this is based on cutting cow numbers in half. Also, work has been cut in half.

Maybe there ought to be a day to go fishing or take a holiday. Kids can change pastures.

Rotational grazing management means selling your grass through your calves. The expense for a 500 cow herd, $20,000. I estimate an additional $200 for keeping a steer 18 months. The average price received per carcass projects at $1,800. The price for 48 calves computes to be $86,000, if they all weigh 600 pounds and if you get that $3.00 a pound after doing your homework and marketing.

Such a transaction cannot be accomplished with the wave of a magic wand. First there has to be a start. It would take 100 calves at $1.50 a pound to achieve the same bottom line, and the market is not likely to permit such a modest windfall.

The operating cost for one-year cows and calves with steers on grass would be $19,000 — $45,000 profit from these 50 cows, about $66,000. I can live on that, and I suspect the average farmer can live on it also. It's a target.

The Dollar Difference

There is a lot of difference between $66,000 and $19,000 via the sale farm backgrounder and feedlot infrastructure, $45,000 in fact for half as much work. And this possibly without even cranking up that John Deere tractor.

The expense of 100 cows, the way the infrastructure figures it, is $25,000. The best you can hope to achieve is $44,000 if you sell at five or six weights. Bottom line the way the above unholy trade plays the game, about $19,000. That won't buy a very upscale automobile. The cowman has to realize that every one of the auctioneers, handlers, backgrounders, truckers, feeders — enlarge that to include inspectors and government gumshoes — makes money. And they are willing to give the producer about $500 for a grown calf. Parity is a no-no term in academia nowadays. It simply means equity.

Equity does not attend transactions when the farmer breeds the cow, calves it out, husbands the animal to sale size — in short, does most of the work. The sale barn is a vulture. It takes a commission on every calf. The order buyer is a worse vulture than the sale barn. If there's a scratch or a barbed wire cut, he'll dock the seller 10 or 15 cents. The backgrounder has a little wheat pasture or forage, usually rented ground locked in the portfolio of some absentee owner. He accounts for some gain, for which reason he works up tall, narrow animals that have to gain. He avoids good calves the way the devil shuns holy water.

The feedlot, sanctified by the university as high technology and agriculture according to the industrial model, is really a disgrace to a civilized society. A cafeteria of feeds is kept before the animal.

Confinement Animals

Confined animals are fed by feed fabricators and salesmen, albeit never a mouthful of grass.

The slaughterhouse has its margin computed and applied. The grocery store makes more for a push across the counter than the rest of the infrastructure put together. The housewife thus becomes the eighth owner of the calf described herein. My computation has it that the average calf travels 1,400 miles from calving to dinner plate.

All the above prompts me to suggest a test, or a model if you will.

Farmer Jones has two pastures, both 400 acres. One pasture has 77 cows weighing 1,300 pounds each on the average. In the second pasture he has 100 cows, 1,000 pounds each. The 77 cows weaned 50 percent of their body weight, thus 650-pound calves. The 100 in the second pasture also weaned 50 percent of their body weight, or 500-pound calves. The 1,200-pound animals weaned off 49,000 pounds. At 89 cents a pound, that's $43,000. The 1,000-pound animals weaned off 50,000 pounds. At 98 cents, this computes to $52,000. The 1,000-pound cows made almost $9,000 more profit than the heavier animals, the same amount of land supporting the gain. The 1,000-pound cow strategy is simply more efficient.

At 1,400 Pounds

A 1,400-pound cow times .03 percent of her body weight equals the amount of dry matter the animal will consume in a day. In a year this equals 14,330 pounds of forage. With a calf at her side, at 250 days she'll consume 17,830 pounds of dry matter, or 17.83 thousand-pound round bales of hay at a cost of $320.94, about 45 percent of that cow's body weight will deliver about $535 in a year. The same general principle applies to the 1,300-pound cow. She and her calf will consume about 16,735 pounds of hay or 16.75 round bales of hay. Each 100-pound less weight subtracts approximately one bale of hay from the equation. Thus a 1,200-pound cow means 15.5 round bales of hay, or $279.60. The 1,100-pound cows at .03 percent consumption, with the above-described herd of 77 will consume 14,545 pounds, or 14.5 round bales, cost $261. With a realized income of $544, the gain becomes at once apparent.

Up and Down

When you come down in body weight, you go up in weaning weight percent. That's simple cowboy arithmetic. It's a given. And this is why genetics govern.

Numbers tell the story:

Expense and Income from Direct Marketing
400 Acres and 50 Cows

Profit, Direct Market 50 Cows = $66,800
Profit, Sale Barn 100 Cows = $19,000

Difference $45,000

Income, 100 Cows Conventional = $44,000
Expense, 100 Cows Conventional = $25,000

Difference $19,000

Owners of 450 and 500-Pound Calves Sold Through Sale Barn

- Farmer
- Sale Barn
- Order Buyer
- Back grounder
- Feed lot
- Slaughter house
- Distribution & Grocery Store
- Housewife

The point in all this should have arrived by now with sledge-hammer impact. Take 100 cows at 1,400 pounds. These animals will consume 17,830 pounds of forage. The 900-pound cows will consume only 12,000 pounds of forage. That is a difference of 547,000 pounds of forage between the 1,400-pound cow and the 900-pound cow. Divide the 547,000 by 12, you can add an extra 44 cows on the same grass. Those extra 44 calves at 90 cents a pound would bring an extra $25,740.

These computations become more intriguing when we remember that I am computing that the 900-pound cows all bring in 65 percent of their body weight, whereas the 1,400-pound cows are bringing in 45 percent of their body weight. The rampant efficiency of the 900-pound cow is at once apparent. The smaller animal does not suffer as much. She spends less time eating to maintain her body condition. She is simply a more efficient cow.

The feedlot system will continue to frustrate cattle producers who are trying to achieve a bottom line. Institutional arrangements that are solidly in place seem to conspire to keep from farmers and ranchers a full understanding of the economics of grass. Grass means forage, perhaps 50 or 60 species so full of nutrients they never can be replaced by feed fabricators, synthetics, medicines and inhumane conditions. For this reason I now ask you to go on a pasture walk. There will be birds chirping their blessings for thirsty ears when shade reaches over grass. Frogs may be goaded to life in the farm pond, but the scene that always takes my breath away is the cow at grass.

The Regulators

Cowboy arithmetic readers go well beyond making calculations of profits to be realized by computing sizes, herds, acres of grass and the market as is. The business of simple arithmetic reaches into every phase of beef production. Some of these additions have been covered in the chapter on pastures. Others are made a matter of record under the heading of *The Mineral Box.*

Now I would like to spend a few notes harvested from more personal experiences.

I have a client in Joshua, Texas, a Santa Gertrudis breeder with whom I've been working for some time. I've been determining which animals have the most tender meat and the best loin, but are also animals that are not extravagant in size. I've done linear measuring. I have used ultrasound. Ultrasound has been used in hospitals to determine the gender of babies. It is also used to view fractures, torn ligaments, and there are practices for doing heart work. In cattle I do what we call *carcass evaluation.* I measure the size of the loin, the intramuscular fat and the amount of body fat each individual animal has. With the birthday and weight of the animal, I can come up with a loin-to-carcass ratio. That ratio needs to be somewhere around 1:3 to 1:5 in order to have a good amount of loin muscle. It takes a set of wide shoulders to support a muscle of that size.

I couple the ultrasound with measurements that make certain the animal is not too tall and framey. I want an animal somewhere between a frame four and five. I want them to be wide in the rump. On cows I want that rump to be at least 40 percent as wide as they are tall. In a bull, the minimum width needs to be 44 percent of the height.

Management

When these minimum standards are allowed to govern, any animal allowed back into the herd will have a lot of meat on it. A cow and a bull with a lot of meat will create an animal with a lot of meat. This makes no reference to EPD. There are standards set by masters of husbandry long before I came along. Under these guidelines, no animal is allowed back into a herd if the females are not ready to breed by the time they are 14 to 16 months of age. None of the bulls are used unless they are ready to breed and settle cows by the time they're the same age.

By following this line of management, puberty in that herd can be kept at an early age. The objective is to achieve maturity at 36 months of age. I do not want animals that keep growing and growing to some ponderous size.

Long and tall and thin animals are easy to stress. They are harder doing, higher maintenance animals. In my opinion, the Santa Gertrudis breed has gone the wrong direction. The breed has a lot to offer, and a few of my clients are starting to make the changes necessary.

Santa Gertrudis hails back to the King Ranch circa 1900. It was created out of the Brahma Shorthorn, perhaps some Hereford. This means a three-way cross. It started from one parent bull. They all came from a bull named Monkey. This stabilized the gene pool. About 1948 or 1949, King Ranch started selling these bulls. Breeding to other kinds of cows followed. In the process some of the genetic potential got lost. The name itself came from a creek that crosses King Ranch, the Santa Gertrudis.

In spite of observations committed to print that suggest degeneration, there are quite a few progressive cattlemen. I have in mind a friend in South Dakota, who has British White cattle. He had been seeking the tallest, longest heifers to go back into his herd. He was actually selecting late-maturing animals. This resulted in retention of heifers with heavier birth weights, and calving problems followed. This problem persisted even though lightweight bulls were being used. Moreover, he was selling the heifers he should have been keeping, and keeping those that should have been sold.

The switch to more balanced, deeper-bodied herd heifers followed our conversations, linear measurements and general analysis. Now he is on the road to restocking the quality of what British Whites are all about.

I have a client at Mogridge, South Dakota. I'm sure he'll excuse me for relating his story. He is a commercial breeder with about 1,000 head of Black Angus type cows and a few Herefords. He's been artificially breeding for 25 years. He buys semen and also buys bulls to clean up with, always high EPD bulls. He created a herd of cattle that were extremely tall and long and thin — very high maintenance animals. During a hard winter he lost large numbers of adult cows. They'd lose their body condition and literally freeze to death.

I met the above-mentioned breeder when he was looking for answers, even for the right questions. I suggested a bull described as optimum in at least three or four points — the wide-shouldered, masculine, wide-rump, short-neck bull. He purchased about 300 straws of semen, a half cc of semen per straw. Laboratory science has learned how to process, say, 25-30 million per straw for syringe insertion, usually at $15 to $20 a straw. It costs about the same to keep a bull to breed a cow.

Using the best bulls available via AI, he is now getting calves far superior to those birthed with reliance on EPD bulls. As I set down these lines, he has decided to keep 20 bulls out of his herd, all at 16 months of age. By the time we get three-quarter blood calves on the ground, bulls will be sold to neighbors. Only superior animals will be kept for herd bulls. A key here is to keep animals in their environment, to avoid excessive transport.

Jim Kelly lives near Rochester, Minnesota. Kelly bought Tom Zimmerman's British Whites, possibly the best British White herd in America. In 21 years, Zimmerman created at least three great bulls. Each has demonstrated the ability to produce superior calves. In Kelly's case I have helped put together the mineral rations needed to support the virility and health superior animals require. Ruthless culling and selection of sire to get the best calves is Kelly's objective and art.

Robert Glenn of Walker, Missouri has about 200 head of commercial British Whites. He uses British White bulls. His herd is in the three-quarter blood range. He feeds out his cattle, averaging 88 percent choice with calves he is producing. Bob was having some *E. coli* scours problems, also retained placentas, eye problems, etc. I answered by evaluating the mineral program. All of those problems were shut down. The conception rate always picks up when minerals are in balance.

Two Clients

Two of my clients in Keokuk, Iowa, Bill Casey and Scott Ensmire, background calves. I helped them develop a herd. Both have some 15 or 20 head, the norm for most of the cattle producers in the U.S. Casey is keeping his frames to 4 to 4.5 — real wide-rumped heifers. He is breeding them to a British White bull. He has intentions of starting a new British White herd.

Scott is a Black Angus breeder. It is always a challenge to build a cowherd that is consistent, that will exhibit a lot of meat, that will be low maintenance, and deliver a calf every 12 months.

Casey and Ensmire and several others have started an organization called Grassland Beef. They are finishing animals on grass, stocking and selling grass beef, thereby answering an ever growing consumer demand for hormone and antibiotic free meat protein.

The reports for this result are endless, and therefore have to be limited. I work with a foundation in New England dealing with rare breeds. New England Heritage Breeds of Richmond, Massachusets, preserves breeds that date back to the 16th and 17th centuries. The North Devon is one breed. They work well as a dual-purpose animal, as do many of the minor breeds the organization seeks to preserve. We have found six or seven cows that fit the profile of outstanding female animals. I am looking for bulls for beef and dairy. The North Devon also does well as an oxen. Needless to say, some of the Conservancy work has to do with hobby people, even dilettantes. The process of finding the correct bull, any breed, is no armchair job.

The Conservancy is also trying to find Herefords without recessive genes. Finding the best sires for Shorthorns and Galoways is particularly challenging. They are not rare breeds, but genetics have gone in the wrong direction so long it is no longer a simple task to reclaim the values now that so many of the bloodlines have been destroyed. The dilution of the Hereford 50 years ago is a good horrible example. Dr. Jan Bonsma said it all and said it well. He was afraid that genetics in America had deteriorated to such a point that it might not be possible to recover without finding genetics in indigenous herds from elsewhere in the world.

The Conservatory

The Conservatory has purchased a slaughter operation and is using organically grown and grass-fed beef. Their market is the high-dollar restaurant complex in New York.

Two things seem to inhibit farmers from marketing their own product. They defend that they are not salesmen. There is only one answer: take a Dale Carnegie course or get out of cattle farming. Others say they simply want to grow cattle, that they have no interest in other work.

Some 80 percent of the cow-calf operators produce 20 to 30 head of cattle. Yet they allow the trade to set up a point through which the production must pass, giving the primary producer very little in return for his creation of value. The trade has also done a number on the consumer. Most of them believe the only source of food is the supermarket. Yet the food shopper who eludes those six steps of commerce by dealing directly with the farmer gets an edible product that does not offend cells and protoplasm.

The Herd

Having the biggest herd in the country is not the answer, not when lower weights, the right dimensions, rapid growth on grass and more direct marketing hold in escrow the key to prosperity unlimited.

Since there are few cattle in New England, the infrastructure has not tainted the minds of remaining farmers. Big organizations cannot operate in such a vacuum.

I have a client in Crozet, Virginia, with 50 cows and an on-scene store where all of the farm's beef is sold. Getting cows pregnant was the problem. Some went as long as six months without working. Using ultrasound and linear measurement tools, I managed to isolate animals fit for replacement. The pregnancy problem evaporated largely because he got rid of open cows, eliminated the gangly tall ones and canceled out the poor-doers. Grass and marketing infuse health to that balance sheet.

There is a fine sprinkling of line breeders and several commercial owners who are making cattle raising work for them. Ted Slanker of Powderly, Texas, and Jim Lents of Indianola, Oklahoma, are two.

Direct Marketing — A Comment

A major meat packer in Oregon posed a question that will continue to haunt feedlots and commodity quality meat protein. Facing cattlemen from across the state, the question came like a rifle shot. "How can meat providers get the quality of meat that will enable competition with direct marketers?" The answer was simply a gale of silence.

The commercial man does not seem to know what is going on in the life of the direct marketers. More available, but ignored, is the fact that the average calf is fed every kind of feed in the world, including protein bypass, roasted manure, silage — and silage does not produce quality taste. The steak taste is flat, not rich. The arithmetic of direct marketing demands its own taste, and excellence.

Chapter 9

The Industry

The movie version of the cattle business leaves little room for in-depth analysis of the modern trade. I have little time for a report on 4-H, FFA or other bankrupt efforts to start youth in agriculture. In the wake of the Farm Act of 1948-1949, youth has been made irrelevant. That measure was fully implemented in the early days of the Eisenhower administration. At that time, the die was cast. Henceforth basic staple commodities — wheat, corn, soybeans, oats, rye, cotton and its seed — all were destined to seek and find the world price level. This happens to be the lowest common denominator in international trade channels. It was the steady lowering of grain prices that enabled the emergence of an enterprise that otherwise could have no reason for being, the gigantic feedlots now marking the plains for airline pilots and polluting the environment for all posterity. Grain priced well below the cost of production — grain producers and their bankers salvaged from bankruptcy with several forms of subsidy payments — was better than grass, this according to scientific truths revealed to the business by college researchers. The physiology of the bovine was ignored, as was the quality of the meat protein produced.

A Moment in Time

Arbitrarily I have to select the moment in time from which to move forward or look back. I now suggest that the big move was hammered into place when it looked like cowmen would demand a price. It was in fact a brilliant fall day at Walden, Colorado in 1962 that the university-spawned idea came to fruition. Starting early that day and working through an October week, cow waddies loaded nearly 90 railcars with feeder cattle for what was to become the largest cattle shipment in 40 years. After some 3,000 critters had been gated aboard, the train rolled east to a 40,000 head feedlot maintained in Minatare, Nebraska. The cattle had been sold from the A-Bar-A property and State Line Ranches operated by Gates Rubber Company. A week earlier some 70 carloads had left Encampment, Wyoming and another 25 carloads had moved out from Northgate. There were other transfers from the cow-calf end of the business to the feedlots. All this complied with the new factory-in-the-field idea being touted by academia, provisioned by politicians and embraced by consumers, all without a perceptible thought of the implications.

The implications had been around for a long time. They didn't have to wait for the Kern County Land Company, with holdings two times larger than the state of Rhode Island, nor did they have to wait for the international holdings to punch up imports. The trade, a few favored packers and chains, knew all about Gates and Kern County and the rest.

Kern County had moved into Arizona in 1948, the year post-war parity met its Waterloo with the Aiken Bill. Basically speaking, the firm's out-of-state ranches were cow-calf operations. On the California side, Kern County maintained stocker ranches and, of course, the great Gosford feedlot. The sprawling giant operated in more than a dozen states and in France, England, Brazil, Australia, Canada, Ireland, and British Honduras. Discovery of oil on Kern County acres in 1936 gave the industrial model for cattle growing and feeding a shot in the arm, sending it on its way. By the time those trains rolled into

Nebraska, the aspiration of all European operators was to make the iron-railed spreads bigger, stuff the cows with high-carbohydrate feeds faster, and time the harvest of carcasses with the likely drop-dead rate of cattle in the pens.

These several developments confirmed on the public the idea that cow-calf operations were gigantic novice-type ranchers and that it took lots of corn to make beef fit to eat. In fact, perhaps 80 percent of the cow-calf operations in the U.S. have no more than 25 head of mother cows in any year. Animals were transferred to sale barns, food producers, truckers, finally to the feedlot and intensive feeding with hormones, coccidiostats, antibiotics and fabricated feeds. Punctuated with shots a-plenty, animals ended up as a commodity-grade product, mule meat in short.

Cows on pasture and the infrastructure that sees them spend the last few months of their lives in a dust or mud-manure choked pen is a few country miles removed from the glory arena of show biz.

The Show

The show ring has dominated the seed stock sector since the days of post-medieval fairs. It has created and turned into current coin the inefficient animal now the prime presence in the commercial sector.

By the late 1960s, show ring judges departed from criteria that named correct body composition the blue ribbon winner. Exotic animals started appearing — oxen types really, but the judges relied on the tape measure to make their judgments. If the animal was taller, bigger, and in compliance with the show ring paradigm, it made the papers, certainly the trade journals and the fancy fair color flyers.

Jim Lents, a Hereford line breeder from Oklahoma, has a word for it. "The beef industry was morphed from a quality business to new gargantuan dimensions." Earlier, the show ring led the Hereford breed into the norm of freaks and dwarfs.

With or without the show ring, no one identified where the industry should go. After the freak and dwarf, the entire industry tried running from the bear. The last few words are Jim Lents terms: "The industry never looked back." That is his assessment, and mine also. Without any move to identify the quality that built the industry, the trade entered the valley of the giants.

Cowmen with savvy had worked with genetics for several lifetimes. In the process, great line breeders developed an efficient animal, cows around 1,000 to 1,100 pounds, bulls 1,800 or 1,900 pounds. This work produced the most efficient, highest quality beef animal on earth. If downsizing was good, the show ring reasoned that rare would be better. The pendulum swung down, and when the reverse came, it swung the other way. The 1930s and early 1940s land of dwarfism prevailed well into the 1950s.

A College Connection

The savvy cowmen who did not have the ultra-small or the humongous animals tried to talk sense to their associates, but they might as well have addressed a fence post. The old prophecy seemed to hold: When the people are ready, the master will come.

The era of the big animal came on with a rush by the 1970s. In the meantime, academia counseled crossbreeding as a way to get the heterotic effect and to boost production. Some old timers called it "the great frame race."

For a good many years it was a joke that all it took to be a show judge was to own a tape measure. Line them up from tallest to shortest — there was your answer. The measuring I speak of and the swirls of hair outlined earlier were unreadable after clippers and combs dolled up the show-ring animal. In any case, I doubt that even one judge knew of such criteria. There arrived the day judgments asked for leaner and trimmer, and the blue ribbon winner had no fat at all. There are a lot of people who still do not realize that the leaner and trimmer beef animal has hit the wall.

It takes more than a blue ribbon to drive the industrial model that has been developed. Quite frankly, the National Western in Denver and the American Royal in Kansas City are more or less an aside nowadays. The infrastructure now drives the model. The feed yard, the packinghouse and the stocker all conspire to hold in place the subsidy-supported cowpen. The lot wants an animal that will grow, grow, grow. The lot wants fast pounds. Quality is bypassed on grounds that the consumer does not know the difference.

The feed yard wants the inefficient, late-maturing animal in order to sell more feed. The packer knows that as a practical matter the smaller animal makes for more efficient operation, but there are more cuts in the taller, bigger steer. "He sits there, the packer, like a buzzard picking bones," more than one cow-calf operator has opined to me. The packer buys at his rigged blue sheet price, using an average that penalizes any animal that exhibits thick short legs, is bred for efficiency, flavor of meat, and economic phenotype.

The Feedlot

Feedlot animals are overfed because the objective is to sell more feed. The model calls for cheap feed. The government obliges by keeping storable commodities at world prices and by saving grain producers from insolvency with deficiency payments. In the final analysis, grain producers are merely a conduit for government checks going to banks and lenders. This charade supports earnings an average of $3 a feedlot bovine, according to trade sources. The fall guy is the cow-calf operator and the seed stock producer.

The animal I talk about in this book is not particularly welcomed at the feedlot. It feeds out in 60 to 90 days. The lot wants animals that'll require feed for 120 to 180 days.

Until the industry figures a way out of the above dilemma, the packers will continue to penalize better quality beef. Thus it seems the sorrier the animal, the greater the premium it draws.

The commercial producer answers all the above by saying, "I sell pounds." He's right, of course. But he does not sell quality. We have to recover the art of producing an animal that will mature and develop on grass, and we can. We were capable at one time. And we can still produce an animal that will be ready for slaughter at 14 or 18 months of age on well-managed grass and forage. This can be accomplished without going to the lots and an unnatural diet of carbohydrates, hormones and antibiotics.

The Diet

There is a razor-thin margin at the feedlot in any case. A change in public policy to protect the American producer economically as well as militarily could cause the feedlot system to evaporate over night. And with animal rights people, environmentalists and consumers on the march, to suggest that the feedlot system on the prairies is a sunset operation is well within reason. *Cattle Facts* computes feedlot profit $3 a head for the past 20 years.

This choice between meat on the rail or in the paddock or pen is difficult to comprehend. The prize 4-H, FFA steer is familiar enough. A several column story anointing the front page of the local fair edition tells the yarn. Some noticeable hands over a sizeable check that factors out to so and so much a pound. Showmanship is escalated when the term "jock" is applied to professional exhibitors. The tradition of the craft permitted them to scrounge the countryside for a likely looking critter at youth shows. At home the animal got the Hollywood treatment. In the lesser ring the jock careered the novices out of business, experience being translated into a win. In my younger years, I'd see them arrive at the fair grounds with the usual two pickup trucks, one for the cattle, a second one for show paraphernalia and feed.

Stuffing the animal full of feed was considered the epitome of showmanship. The point here is that the blue ribbon winner had nothing to do with business arithmetic or economics. Prize money was the lottery of the hour, and the fellow with savvy and a hypodermic needle had the upper hand.

Training at the junior level seemed to demand esoteric moxie on how to make an animal gain weight or lose weight, to stride forward or backup, great importance being attached to feed consumption, how much money can be lost and how clever the boy or girl with the halter becomes. Nutrition, economics and technology were three of four skills youth programs were to teach.

The steer shows that helped unravel the industry over the past 50 years have now graduated into high-level meaninglessness. The subliminal message, translated by show-ring standards, has led to the inefficiency in beef production described in commentary a part of these pages.

In many circles great reverence is attached to the Hereford, Angus, and Shorthorn and people elevate the big time winner as being representative of superiority. The judges often haven't a clue, clippers and curry combs having removed the very signs needed to make a judgment. The model that energizes to guide the seed stock producer and the cow-calf operator is an insurance policy for deficit spending, out-of-balance animals, and progeny that do no make the grade as replacement stock.

The holy writ of the breed associations is the registry. Cattlemen buy bulls and straws of semen based on this paperwork even though sires possessed of suitable prepotency are about as rare as a levitating bull, meaning all four feet several cubits off the ground at the same time.

Encouraging bankruptcy for beef producers is the common legacy of institutionalized blunders.

The National Western

There is a certain thrill that goes with attending the National Western in Denver, but now and then you find someone in the seat next to you who asks, "What has all this to do with nursing calves on a Mississippi farm?"

The commodity beef on conveyor belts out of Iowa Beef is being rejected by ever growing platoons of consumers simply because it is tasteless. It is tasteless because of a genetic elimination of fat, nature's condiment for conferring taste on meat protein. The cholesterol myth is wearing thin as informed consumers discover the spin behind pharmaceutical myths that put mercury in human mouths, fluoride in drinking water, enhancers on meat and drugs into blood streams to repair circuits damaged by errant nonsense in the first place. Qualities such as early maturing on grass, with emphasis on cuts in the high-priced area of the animal are missing in animals that answer the norm of fairs and breed organization specifications.

Conventional Wisdom

It is the conventional wisdom that I call into question with this handbook. Fantasy and myth cannot serve the seed stock producer or the cow-calf operator. The institutional arrangements for taking the product to market is equally errant.

There was a time, some 30 years ago, when ranchers' chest thumped their independence. They were free of the restrictions that attended row crop production of grain and therefore did not have to endure the humiliation of government checks in order to survive. A measure of economic speculation has now filtered down or crawled upward via capillary action — I don't know which—and put the public policy in perspective.

The old rural saying that you can detect Spring in those places, the voices of birds, the buds of trees and the glands of men, has now been supplemented by the obscenity of the latest farm bill, cycled every five years. Eisenhower called corn, wheat, soybeans, etc., "those political crops." They were and are.

Starting with the Truman administration, it has been public policy to bring grain prices down to a world price level. This much accomplished, it was discovered that giant combines, fertilizer efficiency, and super-human management wasn't enough to keep mega-farms in business. They needed land banks, deficiency payments — in a word subsidies — so they could pay their notes at their bank and permit grain to flow into trade channels at world prices. Cheap grain in turn enabled a new science, the feedlot.

All the above was attained by the concentration of agribusiness, farm crises one after the other, rural bankruptcies, and — not last — a virtual cessation of cows being finished largely on grass. In any year, the largest grain growers receive about 84 percent of the payments that run in the neighborhood of $30 billion a year. A dribble goes to 1.5 million smaller family farmers.

The subsidy system has its incentives, the feedlot. Corn that added the cost of subsidies to the price at the tailgate would visit on the feedlot system its legacy as a sunset operation. But the public policy called North America Free Trade Agreement (NAFTA) and World Trade Organization (WTO) figure it is a small price to pay for out-migration of U.S. industry to short-change small farmers, especially cattlemen.

Accordingly, there is no enforcement of Packers and Stockyards Act of 1921, no enforcement of the principle that brought down the hammer of anti-trust on major packers by the consent decree of 1922.

Grass Fed Beef

Grass fed meat has more omega-3-fatty acids and fewer omega 6, the last being a heart disease culprit. Grass fed beef contains beta-carotene and CLA, a beneficial fat. By way of contrast, it is being revealed by disciplined research that the human being is not best served by grain fed beef, much as the cow is being abused by being stuffed with corn and urea, all this without emphasis on hormones, Rumensin, and alchemy from the devil's pantry. The fact that feedlot cattle could not survive an additional six months in the pens speaks for itself.

Primary Producers

Cow-calf operators, backgrounders, feeders, all know that the system is wrecking prices and making it difficult, if not impossible, for the primary producer to enjoy solvency. Implicated are the big meat packers and retailers that have conspired or at least worked in alliance at the business of controlling the market. It does not take four fingers to count the packers that control over 80 percent of the slaughter animals in the United States — IBP (Iowa Beef Packers) leading the pack unless they merge and change their name once more. IBP is a partner with Wal-Mart. Wal-Mart is the largest corporation in the United States, having come from nowhere via socialization of investment. These corporate panhandlers have convinced almost every town, city, state to excuse them from taxes, give them land and utilities, and now they're selling meat with payment deferred pending actual ring out at the checkout scanner.

This arrangement translates its power into downward cattle prices. IBP bids just low enough and high enough to keep the traffic coming without drying up production. With no antitrust enforcement under the auspices of the last ten administrations, the losses to cowmen have been staggering, $400 per animal during the last few years alone, this the share of the consumer dollar the producer should have enjoyed.

Fast food chains are a great part of this confiscatory equation. They reject doing business with small packers the way Fido rejects a mock hamburger. In fact, only the biggest of the big can supply the demand volume.

The yellow sheet of the 1960s and 1970s is now called the blue sheet. The blue sheet carries what the packers want to report, which means wired or rigged data. This means lower prices are reported than those paid in order to create the atmosphere for still lower prices. In any case, much of the role is well removed from USDA data. At the packer and retailer level, the deals are what we used to call "brother-in-law," or "special favor." Negotiation on prices is not a factor. IBP publishes a price. The

Attorney General and the Secretary of Agriculture both defend that the Packers and Stockyards Act no longer applies by common consent.

This situation calls into question federal statistics, the deficiency payments for feedlot grain suppliers and the parity computations that are published in fine print only because statute law requires it.

World Trade

The misfeasance and malfeasance mentioned above is furbished and refurbished by the World Trade Organization. We are asked to believe that a strong local community is less important than filling the American malls with cheap foreign goods. A safe, local source of food has been pronounced anathema by international hustlers who find a profit in sourcing the poverty pockets of the world for sales in the high market, closing out domestic producers in the process. Red meat production is not exempt from this objective. In fact one of the fast food chains has answered the objective of buying only foreign meat. Domination of the food supply by a few companies has been largely accomplished. The global market argument would be rejected by even high school pupils except for the atmosphere of intimidation under which classroom instruction commonly proceeds. Every child knows the farmer does not sell in a global market. The farmer sells locally. In most cases there is only one place to sell, and this is on a "what'll you give me?" basis. There certainly is no competition between the USDA's parrots.

In 1922 a consent decree closed down monopoly at a time when five packers had achieved a 40 percent market domination, all this in response to the Packers and Stockyards Act. Today 80 percent domination draws a yawn from the Justice Department and USDA. Jefferson's vision of broad-spectrum distribution of land and money seems to have died to the degree government has been sold to the highest bidders called campaign contributors. I do not think I need to belabor the point that great corporations with U.S. addresses now import beef, lamb, even chickens. Meat packers now import live cattle — commodity beef, that is — boxed beef and pork from around the world, even though this puts our people at risk from a food safety point of view and even though it closes down businesses, producers, ranchers and farmers. If the cow is an unpaid worker, as are the microbes in the soil, it still cannot compete with slave labor. As the economy is destroyed, most people look the other way. It is a heavy burden, this knowledge that great corporations with U.S. addresses are destroying the very basis for economics and of industry, the farm, the ranch and the raw materials producers of oil, cattle or grain. Destruction of such scope is a national security issue. It is as threatening as bombs dropped on New York.

Breeding cattle and producing a product with taste arises to the dimensions expressed in the last few paragraphs. It will take superior flavor to recapture the local market and to annihilate that seven stage ownership of a calf, those tons of rubber left on ruined highways as trucks move cattle across the states first to feed them out, then back to slaughter or market. These uneconomic practices are subtracted from the earnings of the primary producer.

An Observation

It seems to me that cattlemen, no less than the general population, are being misled on the benefits of the stock market. Wall Street is viewed as the economy. In fact, it merely represents the global corporations that have held in thrall farmer, every cattleman, every shift worker and all the supporting personnel who make the economy a working reality. They roam the earth for every scoop of real wealth and covet every spit of land and all the raw materials taken from Mother Earth, starting with grass and taking in all the elements on the Mendeleyev chart. Yet we all are told the stock market is the economy by CNN and the public prints.

Global free trade and fast track have little to do with the cowman except to drive prices down to the level of the lowest producer said to contribute to trade channels. In fact, fast track and Trade Authority merely give the government authority to trade for the best interest of global corporations, Cargill, Con-Agra, Tyson, Iowa Beef and ADM included.

All support technology that is inappropriate — genetic engineering, killer technology to battle fungal, bacterial and insect crop destroyers, hormones, Rumensin and, not least, soy.

My co-author tells me that soy has been implicated in the sterility of test animals, and there is no reason to not suspect the same for bulls. ADM taunts TV viewers with the proposition that pastures should be plowed under to make way for soybean production and its superior efficiency of protein production. The implication is that farmers will do better producing soybeans for ADM than they can grazing cattle on grass. Well-funded scientists speak as with one voice as they chant the praises of soy for what ails mankind. Each statement is well credentialed by United Soybean Board, American Soybean Association, Monsanto, Protein Technologies International, Central Soya, Cargill Foods, Personal Products Company, SoyLife, Whitehall-Robins Healthcare and the Soybean Councils of Illinois, Indiana, Kentucky, Michigan, Minnesota, Nebraska, Ohio and South Dakota.

The major objective is to gain acceptance of soy milk and ice cream and soy foods for human beings. In 1980, soy milk merely nibbled at dairy with $3 million in sales. By the end of the 20th century, soy milk sales were well in excess of $300 million.

$300 Million

Of interest here are the estrogen-like flavones found in soybeans. The estrogen part of the soybean used to be discarded as a waste product. This no longer happens, not with some 72 million acres in production and pens full of captive bovines starved into eating the stuff.

Soy is now sold to health food stores, not as the poverty food it is. The four color ads say it will prevent heart disease, cancer, and annihilate hot flashes, build bones, and confer perpetual growth on meat, milk, butter and eggs for *Homo sapiens* consumption. Butter and milk have been demonized and condemned, not only by the aforementioned agencies, but by USDA and other government bodies as well.

Vegetarians have embraced soy foods with eager-eyed ardency of Harpo Marx chasing blondes. Admittedly, food scientists can make peeling paint taste like ice cream. This does not change the nutrient value.

The soybean contains toxins or anti-nutrients, enzyme inhibitors that block the action of trypsin as well as enzymes required for protein digestion. Encased protein in the bean's protein-makeup produce extreme gastric distress. The pancreas suffers and cancer often is provoked in the animal. The bean also contains hemoglutin, which promotes red blood cells to clot.

Ranchers tell us soy contains goitrogens. These compromise thyroid function and the ability to metabolize sugars in corn. More sinister are the phytates. These block the uptake of minerals — calcium, magnesium, iron, and zinc.

Industry answers most of the above with high-temperature processing. This results in a denatured product, for which reason all animals fed soy need lysine supplementation. Feeding experiments have revealed the necessity for increased vitamin E, K, D and B-12 with deficiency symptoms for calcium, magnesium, manganese, molybdenum, copper, iron and zinc.

An industry content to feed chicken litter, reprocessed manure, corn and soybeans cannot be expected to take notice because, as one feedlot operator put it, "They'll eat it before they eat each other." The breeder of seed stock might seek the edge, so to speak, the way a baseball hitter learns a fine point in hitting.

Sally Fallon and her co-author, Mary G. Enig, write in *Tragedy and Hype* that celibate monks living on a soy diet find the food quite helpful since it dampens libido. Dampened libido in a monk is one thing, the same effect in a sire is another.

The luckless hunt for suitable sires on which to breed and redeem the proud breeder's heritage once the property of cowmen seems compromised on many fronts. Faltering libido can be computed with simple arithmetic. A 100-gram fix of soy contains the estrogen equivalent of a birth control pill. Reproductive compatibility has been observed for many species, sheep and pigs included. Soy meal blocks calcium and causes vitamin D deficiency.

Arithmetic tells us that soy infant formula delivers the equivalent of at least five birth control pills a day. There is no such effect from consumption of human or cow's milk.

It has been speculated that sexual orientation—human and animal—is determined by early hormonal environment.

The bovine research that answers these perplexing questions has not been accomplished. The international companies are not interested, but the cowmen should be interested. In the final analysis, it is up to him or her to stay alert, find answers.

As I set down these lines, the industry continues to be guided by the absconders. Yet we know, it costs only nine cents to put a pound of gain on a steer in New Zealand. It costs 35 cents to add a pound of gain to a grass fed steer in the United States. This suggests that there is no real basis for unregulated free trade. Brazil can ship to the United States at less than half the U.S. break-even point.

I have added these few notes to show how technology, global trade and the high-level scam mold one's destiny. It is no accident that an international company named Siza with a French address now controls 40 percent of American dairy, the so-called organic Horizon dairy included. This one firm is well into the business of canceling out dairy and converting most of the market world to soy milk. The bovine producer alone can do something about it. That is what this report hopes to accomplish.

Afterword

I now abandon "I," Gearld Fry as narrator, and hand the text without intervention back to "I," Charles Walters, for an afterword.

As a child of the Dust Bowl, I have memories that are as fresh today as when tons of dust closed down the sky and smothered the pastures. One is an old *Collier's* article, "Land Where our Children Died." It was this article that coaxed my parents to migrate out of western Kansas in the mid-1930s. Many farmers hung on, certain that rain would come, that the cycles of life would turn those wonderful acres into greenery. I did not want our family to move, but there was my oldest brother forever reading and reciting that article.

Later, in the smelter-scarred ex-industrial town of Iola, Kansas, I became converted to my brother's and my parents' view the day I saw a short subject at the movie theater where I worked. It came styled, *The Plow that Broke the Range.* It was published by the Resettlement Administration and carried the imprimatur of Rexford Tugwell an assistant secretary of agriculture in 1933. I have been told that Tugwell used it to convince

Congress that something should be done about the rape of the high plains for the purpose of producing a surplus of wheat. I had to use that film during the early days of *Acres U.S.A.* to get farmers even to listen to the eco-farming story. I have heard it referred to as a work of art on par with the still photography of Arthur Rothstein and Dorothea Lange. I can see that opening scene even now on a rolling sea of grass and hear the cultivated English voice of a commentator who explained how 40 million acres of high plains grassland anointed the landscape before the arrival of sodbusters. The music-fortified scene was tranquil, an effect later used by Hollywood to announce Conrad Richter's *Sea of Grass,* starring Spencer Tracy.

As the plowman arrived, the music grew louder. Close-ups showed grass plows rolling amid furrow after furrow, and with music usually reserved for trouble in the streets, wheat began to grow. The discerning eye could see puffs of dirt rise toward the sky.

Time's passage was depicted with a World War I headline about soaring wheat prices. Now you could see tractors marching across the former prairie like members of the Coldstream Guard breaking sod. The graphics were intoxicating, mesmerizing. A stock ticker illustrated the retreat from reality. In those days film-makers didn't morph a scene. They allowed the screen to go black. This lasted a second or two. Then the skull of a bovine peered from under a bank of dust. A bowl almost covered by dust set up the announcement, "40 Million Acres of Plains Totally Ruined by the Plow, 20 Million Acres Badly Damaged."

The film ended on that note. It asked, *What is America going to do about it?*

I wanted these early subscribers to ask why grass was gone from the high plains, why the plow had its deficits. Discussions, when they flowed from these meetings and speeches, led to the heart and soul of correct agriculture.

With global warming forever a news topic and powerful ad dollars instructing farmers to turn pastures into row crop acres for greater protein production, it seems appropriate to end this volume with a short piece on topsoil erosion, water utilization, air pollution, water pollution, and the conditions governing plant growth, emphasis grass, plant physiology, and the natural nitrogen cycle and carbon cycle.

Much of what follows was prescribed in two books I co-authored with Leonard Ridzon, *The Carbon Connection* and *The Carbon Cycle.* Both pointed to the quadruple destabilizing factors of our time — topsoil erosion, water shortages, air pollution, and water pollution. The role of *Bos taurus* and *Bos indicus* maintained on grass in dealing with the above should be at once apparent. The carbon cycle simply has to be the solution for it is the key factor in producing cattle as nature intended.

Carbon is the second most abundant element in a plant. Water is first. Plants harvest their carbon out of carbon dioxide in the atmosphere. Plants take in this jewel of an element through the stomata, pores on the underside of leaves. The higher the carbon dioxide concentration, the better the plant grows. Indeed, plant growth improves until the carbon dioxide concentration exceeds 2,700 ppm. Geologists tell us that the CO_2 concentration in the air 120 years ago was 275 ppm. It is now about 380 ppm.

The Michigan scientist I referred to in an earlier chapter, William Cooper, has best explained the commonly accepted theory that all fossil fuels that were and still are underground were once plant life. This means all carbon dioxide in fossil fuels was originally the CO_2 in the atmosphere.

Now consider this: the stomata on leaves open and close as ordered by the physiology of each species. The stomata opens to admit air in order to take in carbon dioxide and trace plant nutrients. Dan Carlson, an innovative plant physiologist, has held and proved with experiments that sound waves from birds give the plants an assist during the hours of morning mist for which reason birds broadcast a veritable concerto while the rising sun starts to wake up the rotating planet.

The leaf in the presence of light separate the carbon out of the CO_2 and process it into energy and release the oxygen back into the air — photosynthesis.

Plants require moist soil for growth, yet nature has decreed that the stomata will pour out fully 99 percent of the moisture that the roots so artfully take from the soil. Obviously, a high concentration of CO_2 in the area around leaves makes it unnecessary for the stomata to stay open very long. Closed stomata lower transpiration to conserve soil moisture. Again, this is a variable according to species. The Chilean tamerago, for instance, thrives in areas that do not see rainfall for ten years at a stretch. They capture moisture blown in from the sea as humidity and feed it to the soil. Pasture grasses and herbs are not tamerago plants. The anatomy of their life style is described as follows:

The longer the stomata stays closed, the less water a plant takes from the soil. The difference between a desert plant and a rotation pasture is that grass increases the tempo of capturing water from the soil releasing it into the air.

As roots absorb water from the soil, osmosis pulls soil water toward the roots. This soil water contains dissolved nutrients. The trail of nutrients, tracked along with microorganisms was described in the early years of the last century. The microscope window revealed the fact that available nutrients are much smaller than the micron size considered suitable for supplement feeding of livestock.

Rich soil has more suitably sized nutrients than less fertile counterparts. The difference is seated in decaying organic matter. Nutrient loads delivered by microorganisms enter into soil water. Nutrients in water as well as in the bodies of microorganisms permit plants to thrive with less water.

When moisture and temperature are maximum for plant growth, they are also maximum for macro and micro life forms to proliferate and accomplish the chore of plant residue decay.

Decaying materials give off carbon dioxide much as human beings do when they exhale. CO_2 is heavier than air. It tends to concentrate near the soil surface. It hovers there for a while, then obeys the law of diffusion much as smoke from the *laissez-faire* section of a restaurant invades the non-smoking section. CO_2 moves off the soil's surface, finally feeding the plant leaves. The leaves in turn take in a suitable diet as explained earlier.

Clay soils are at a disadvantage in the fertility game. They absorb water poorly. Sandy soil unadorned with organic matter also does poorly. It cannot hold water. Either can't or won't means a starved plant.

Organic matter in any soil will hold more water. An organic mulch will preserve water and perform well in collecting water from rainfall without the runoff usually associated with clay. Mulch also permits release of carbon dioxide and allows the plant to require less water for growth.

Moreover, mulch stops wind erosion and limits evaporation. In appropriate situations, a mulch can serve as a weed barrier, and is therefore a prime justification for no-till and low-till on row crop acres. When fertilizers are used, mulch holds them in place.

Organic Matter

These several considerations help explain the mandate for grass cover for most of America's acres and the primacy of cattle in maintaining that area. Turf, organic matter, animal droppings, all figure in the equation.

All of the above conspire to bind sandy soil together for the purpose of greatly increasing the moisture level capacity for extra mileage out of rain. Organic matter increases the carbon exchange capacity to prevent nutrient loss via leaching. Organic matter loosens clay soil thereby enhancing water absorption, root penetration and CO_2, oxygen exchange.

Organic matter helps govern pH, allowing plants to tolerate deviations from the norm. Organic matter performs its supreme function often by feeding microbes capable of refining nutrients, detoxifying pesticides and much of the toxic genetic chemistry now polluting agriculture.

I mentioned air pollution as one of four troublesome consequences of industrial agriculture and over-industrialization of planet Earth. A few afterword words are therefore in order. It's a disturbing fact that atmospheric carbon dioxide has increased over 100 parts per million over the past 100 years. Plants have responded with more bins and bushels, but the scientific world warns that, in excess, carbon dioxide is pollution and have assigned to it the spectrum of global warming. This global warming can be negated by building the organic content of farmland by one-tenth of one percent each year. The scientist who made the computation wasn't sure how this could be done. It certainly couldn't be done by converting pasture to soybean ground for a touted increase in protein production. Conversion of more row crop acres to pastures would be a help.

It is now computed that the soil organic content of farmland is down 20 percent of what it was pre-settlement. This poses a question. Did the current excess come from burning fossil fuels or could it have come from agriculture's excess conversion to the factory-in-the-field industrial model? Agricultural Research Service scientists, sponsored by the USDA, now realize that the type of no-till represented by grass-fed cattle on permanent pasture permits the routine accumulation of that one-tenth of one percent increase in organic matter. Some scientists hold that if we compared the excess carbon in the air and the missing carbon from farmlands, they would be equal.

This brings us to renewed consideration of Edward Faulkner's *Plowman's Folly.* We can hardly reject the thesis that plowing destroys many soil life forms required for healthy root growth and preservation of miles of fine root hairs. Carbon, soil life, and root hairs centered in the soil are oxidized to carbon dioxide, and the plowed-under plant is not there to capture it.

Most of the water problem could be solved with attention to the corn connection.

The Corn Connection

Putting a cow on a corn-heavy diet may be only slightly less damaging than feeding soybeans. Corn is an errant member of the grass family, however it has not evolved in harmony with the need of the herbivore. In the feedlot and on some farms, it inserts itself into the diet and muscles of animals with reckless abandon and endurance. Controlled starvation will cause a ruminant to consume corn, albeit not without stress that inoculates low-level infection and demands antibiotics. Feedlot bloat is not alone in tormenting feedlot animals or cows on some farms. It is being argued that the general switch to corn sweeteners circa 1980s authored the startup of adult onset diabetes in epidemic proportions.

This switch from glucose to fructose for man and animal has consequences that run beyond instant comprehension. This sugar is metabolized quite differently than the glucose in grass. A mantra about the high levels of saturated fats in feedlot beef hardly needs exposition in these paragraphs. Suffice it to say that over 80 million acres of corn ought to be returned to pasture to a large extent. University folks have computed that each bushel of corn consumes a half-gallon of fossil fuel energy as pesticides, nitrogen, herbicides, the drainage from which rolls down the Mississippi to pollute an area the size of Rhode Island in the Gulf of Mexico. All this to deliver excess carbohydrates and linoleic acids into the diet of the consumer. Cowmen who feed grass are not benefitted by that $190 billion farm subsidy program, but cowpen keepers are, and they use their cheap corn largess to beat down the cow-calf and cowherd farmer, the foundation supplier.

Holistic Resource Management

There is no shortage of knowledge. A good move would be to maintain a lush sward of grass on all rangeland as taught by Holistic Resource Management. Grass is the champion plant for the purpose of squeezing carbon from the air. HRM stands in stark contrast to practices endorsed and supported by public policy.

Soil in the ranching country that straddles the Ogallala Aquifer should be largely in grass, this according to nature's design. Yet the Ogallala Aquifer under the Texas panhandle is being pumped dry by growers of corn. Some are producing up to 280 bushels an acre with constant flooding of the soil's surface. Industrial model agriculture says keep it wet and keep the fertilizer on it, never mind destruction of the soil. All this to grow corn that has no suitable market.

One operator with whom I have a speaking acquaintance has 34 center pivots capable of covering most of a section. It costs more to grow this corn than the market can pay. Government payments pick up the slack. If irrigation is interrupted or unavailable, then crop insurance pays for a crop that should never have been planted in that area.

The subsidy train rides along. Its passengers are the feedlots, the poultry pluckers, the confinement hog operators, Cargill, ADM, the lenders, the farmer being a conduit for money to reach the above.

The owner dilemma is the same that has faced all family farmers. Moderately sized business has the option of moving to poverty pockets of the world for cheap labor or going broke. Small business has no option, not with town dads subsidizing Wal-Mart with tax increment financing. The cow-calf producer can survive only by getting smarter, not bigger.

While viewing the distribution allowed by public policy, I am nevertheless required to observe a counterpart, especially in ranch country. I have in mind several grazers who follow the HRM text. Malcolm Beck of San Antonio, Texas, my eyes and ears in reporting much of agriculture, visits HRM ranches. "They have thick grass all over the place. No irrigation, no fertilization, no supplemental feeding, no sprays, no wormers, no hormones. They just grow naturally. They are the healthiest animals you ever looked at. Using rotational grazing, they have it to a science of grazing animals. Beck reported, "With many ranches going broke, these groups are making money, with an average cost of $200 getting an animal to market in range country, some good grass farmers have only an $85 cost," this with grass, no corn. When a pasture is ready to harvest, they turn the cows into it. Without routine harvest, all carbon becomes tied up in grass on top of the soil, an invitation for lightning to strike and a burn-off. "The only problem these ranchers have is a shortage of animals."

Hopefully, Gearld Fry's treatment of the subject has provided an assist.

— *Charles Walters*

Index

Eco-Farm: An Acres U.S.A. Primer

BY CHARLES WALTERS

In this book, eco-agriculture is explained — from the tiniest molecular building blocks to managing the soil — in terminology that not only makes the subject easy to learn, but vibrantly alive. Sections on NP&K, cation exchange capacity, composting, Brix, soil life, and more! *Eco-Farm* truly delivers a complete education in soils, crops, and weed and insect control. This should be the first book read by everyone beginning in eco-agriculture . . . and the most shop-worn book on the shelf of the most experienced. *Softcover, 476 pages. ISBN 0-911311-74-2*

Weeds: Control Without Poisons

BY CHARLES WALTERS

For a thorough understanding of the conditions that produce certain weeds, you simply can't find a better source than this one — certainly not one as entertaining, as full of anecdotes and homespun common sense. It contains a lifetime of collected wisdom that teaches us how to understand and thereby control the growth of countless weed species, as well as why there is an absolute necessity for a more holistic, eco-centered perspective in agriculture today. Contains specifics on a hundred weeds, why they grow, what soil conditions spur them on or stop them, what they say about your soil, and how to control them without the obscene presence of poisons, all cross-referenced by scientific and various common names, and a new pictorial glossary. *Softcover, 352 pages. ISBN 0-911311-58-0*

The Biological Farmer

A Complete Guide to the Sustainable & Profitable Biological System of Farming

BY GARY F. ZIMMER

Biological farmers work with nature, feeding soil life, balancing soil minerals, and tilling soils with a purpose. The methods they apply involve a unique system of beliefs, observations and guidelines that result in increased production and profit. This practical how-to guide elucidates their methods and will help you make farming fun and profitable. *The Biological Farmer* is the farming consultant's bible. It schools the interested grower in methods of maintaining a balanced, healthy soil that promises greater productivity at lower costs, and it covers some of the pitfalls of conventional farming practices. Zimmer knows how to make responsible farming work. His extensive knowledge of biological farming and consulting experience come through in this complete, practical guide to making farming fun and profitable. *Softcover, 352 pages. ISBN 0-911311-62-9*